The Story of MOORE BROTHERS KELVEDON, ESSEX

by

Bryan Everitt

In memory of my father
HUBERT EVERITT

Reprinted March 2000

ISBN 0 9534701 0 5

Published by Bryan Everitt, Chelmsford, Essex
Printed by The Lavenham Press, Lavenham, Suffolk

Contents

Index of Illustrations

Pictures

Figures

Colour Pictures

Appendices

Reference Sources

Albion And Crossley Buses in Camera — S. J. Brown.
Bedford Vol. 1 — S. Broatch and A. Townsin.
Duple. A Pictorial Survey — Eric Ogden.
Eastern National Fleet History (Pt. 3 - 1960 to 1984) — published jointly by the PSV Circle and The Omnibus Society.
Economic History Review 2nd Series Vol. 29 (1976).
Feathers In Their Cap — R. N. Hannay.
Hatfield Peverel Overseers Accounts (1804 - 1817).
Horse-drawn Carriages — John Thompson
Just The Ticket. — David Thornton.
London's Buses Vol. 2 — Blacker, Lunn, Westgate.
The Bell Inn Feering — S. Jarvis.
The Best Of British Buses (No. 1 Leyland Titans 1927 - 1942) — A. Townsin.
The Best Of British Buses (No. 7 AEC Regents 1929 - 1942) — A. Townsin.
The Best Of British Buses (No. 8 The Utilities) — A. Townsin.
The Rise And Rise Of Road Transport 1700 - 1990 — Barker and Gerhold.
The Years Between Vol. 1 — Crawley, MacGregor, Simpson.
Thirteenth Report of the Commissioners of Inland Revenue (1869 - 1870).
Thomas Clarkson (1864 - 1933) — Chelmsford Museum Service.
Various editions of Pigot, Kelly, Benham and Post Office Directories (1825 - 1945).

Reference has been made to a number of periodicals including *Bus & Coach, The Transport Journal, Motor Transport Year Books, Chevrolet & Bedford Magazine, Buses* by Ian Allan Publishing.

Research has been carried out at The House of Lords Record Office, The Public Records Office, The Newspaper Library, The Patent Office, The Family Record Centre, The Essex Records Offices in Chelmsford and Colchester, Companies House, The Traffic Commissioners Office Cambridge, The Omnibus Society Library, Coalbrookdale and Kelvedon and Feering Local History Museum.

Introduction

This book is being published to coincide with the two hundred and tenth anniversary of the birth of the founder, Edward Moore, in November 1788.

Edward Moore was an honest and fair man but with an eye for business. These characteristics were to be inherited by four generations of his family, which in no small measure contributed to the success of the company which lasted for almost one hundred and fifty years.

Compiling this story has brought me immense pleasure which could never have been achieved without the help of the Moore family, ex-employees, and many other people.

I must mention in particular, Patricia Bowdidge (née Moore), Mollie Moore, the late Doris Moore, Dudley Moore, Renée Moore, Stanley Moore, Eddie Warner, Peter Weinrich, George Buck (jnr.), Charlie Lovick, Cyril Eady, Ron Parker, Cyril Mayhew, the late Doreen Pitchford, Geoff Dodson, Rev. David Thornton, Mr and Mrs Arthur Frost, Joan Dale, Brian and Arnold Brown, Peter Snell, David Edwards, Dave Jones, Cyril Tebbell, Barbara Whittaker, Derek Brown, Ken Thomas, Derek Bacon, David Ricks, Miss Myfanwy Garrard, David Bevan. I have received valuable help from Omnibus Society members, and in particular Reg Westgate, Frank Simpson, Alan Mills and Pat Lidgett.

Compilation of the fleet history and disposal listing was only made possible through information provided by John Taylor and Geoff Mills. Roger Atkinson of the Transport Ticket Society was very helpful in identifying and dating early fare tickets. My thanks also to Peter Allsopp of the Traffic Commissioners who made my visits to Cambridge both informative and pleasurable. Many visits have been made to the Essex Record Office in both Chelmsford and Colchester, the Colchester Museum and Local Studies Library, where in all establishments friendly help has never been lacking. The enthusiastic cooperation from Marks and Spencer plc, Complete Photographic Services, Prontaprint and T. Cribb and Sons is greatly appreciated.

To all the above and to others, not mentioned, who have helped in so many different ways, my sincere and grateful thanks.

Last but by no means least this book would not have been possible without the continued support of John Scott, Stanley Wood and my wife, Kathleen.

John was able to take my 'stone-age' script and transcribe it into the computer technology of today. With constant changes to the manuscript being necessary as more material became available, this was an invaluable aid and the prospect of writing this book without John and his computer just doesn't bear thinking about.

Stanley did his best to keep my English on track and made many helpful suggestions in telling the story.

And for Kathleen during the last five years it must have seemed at times as though there was nothing else to life but buses. For her patience, understanding and help I will be eternally grateful.

To everyone, thank you very much.

Bryan Everitt.
November 1998

MOORE BROS. - THE OWNER'S FAMILY TREE

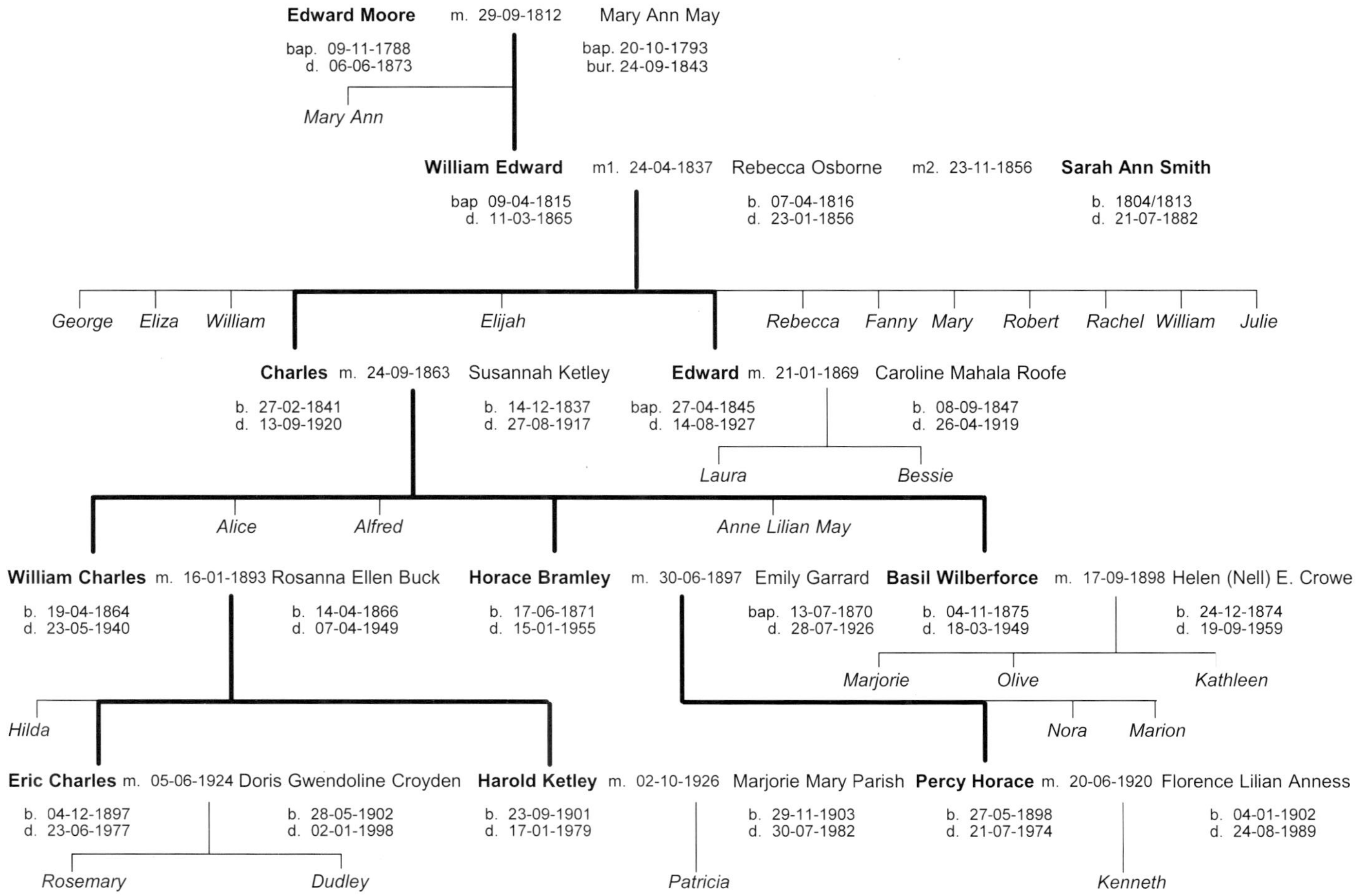

1 An impression from 1935 of Edward Moore's carriers cart.

2 The Bell Inn, Feering.

Chapter 1

Inspiration

The year 1815, as well as being momentous in British history when Wellington defeated Napoleon at Waterloo, was also an historic year in the annals of the Moore family for it is said that on 20th February Edward Moore started his carriers business. So who was Edward Moore, this 'Pioneer of Transport' as later generations of the family were to proclaim him?

Edward was born in the latter part of 1788 and was baptised on 4th November in the Parish Church of St. Andrews in Hatfield Peverel. He was the third child of Alice and Edward Moore and in a family of nine he was fortunate to survive since a number of his brothers and sisters did not even reach maturity — a sad reflection of those times when infant mortality was very high.

Like many other workers in this part of Essex, Edward's father found employment on the land as a farm labourer, as did his grandfather before him, but with little financial reward. The young boy was destined to follow in their footsteps, and clearly young Edward witnessed the family's struggle for survival under harsh conditions. More than once his parents had to apply to the Parish for Poor Relief. We see that in 1804 two shillings and sixpence was approved for the purchase of a pair of shoes for his mother. This must have left a lasting impression on the young Edward and a determination that he would not suffer in the same way.

We know Edward moved from Hatfield Peverel to Feering in the early 1800's but what prompted him to do so is uncertain. At the time Edward was farm labouring so perhaps he perceived there were better employment opportunities where the seed-growing industry was flourishing. More romantically, maybe he could not resist the lure of Mary Ann May whom he married at Feering All Saints Parish church on 29th September 1812. Edward had now reached the age of twenty four. The following year saw the arrival of their first child who was christened Mary Ann on 18th April 1813, but sadly Mary Ann died in March 1823 just before her 10th birthday.

Just a month after the start of the carriers business Mary and Edward's second child was born. William Edward, who was later to take over the business, was baptised on 9th April 1815 in the local church of Feering — there were to be no further children.

Whatever the reason for young Edward's move he was soon slipping into his new life, and in 1815 the carriers business was founded which as we have already seen earned Edward Moore the title 'Pioneer of Transport'. To some this title may seem strange since for many centuries there had been the wains and waggons which were always prepared to carry passengers provided space was available. More familiar to many would be the stage and mail coaches which criss-crossed the country, running to published schedules from the 1600's. With the Turnpike Act of 1695 the roads of Britain began to improve such that the speeds of coaches could increase and so gain greater popularity with the travelling public. In fact it was only the arrival of the railways during the 1800's that forced their decline and finally their demise by the end of the century.

But who were the 'travelling public' that used the services of the stage and post coaches? To travel by coach was very expensive and therefore a domain of the rich. Furthermore, the journey had to be pre-booked with fixed stopping points at a post house or inn. For the majority who could not afford to use these services it was a matter of walking, even where great distances were involved. In fact getting from place to place by walking was far more extensive than many people realise. Therefore, bearing in mind we had these forms of transport, what earned Edward Moore the title 'Pioneer of Transport' thereby making his business unique and setting him apart from the rest?

Although the prime role of the carrier was the conveyance of goods, as it is today, it would nonetheless pick up and set down passengers along the route. Moreover, stopping places were not predetermined nor did the journeys have to be pre-booked. The carrier operated to a timetable and therefore the local carriers cart can be regarded as the predecessor of the stage carriage bus of today with Edward Moore making an early entrance into this mode of transport with a business that succeeded and survived for almost 150 years. To support this view we find Moore's publicity of the 1930's proclaiming that they were the 'Oldest Established Operators of Public Transport in the Eastern Traffic Area'. We can also see from articles published in professional transport magazines that the company has been given the distinction of being 'the oldest family undertaking in the passenger vehicle industry'.

Edward Moore's carrier was a simple bone-shaking cart and hauled by a single horse. We do not have a picture of this cart, but a cart entered by the Moore family in the Silver Jubilee procession of 1935 was intended to show how this may have looked, see Pic.1. It is said that Edward used a cart instead of a van because of his opposition to a tax levy of £8.8s.0d. for road springs, which had to be fitted to a van. It is interesting that this tax was repealed in 1825 by Act of Parliament, but the tax had only applied to non-metallic springs, so his remarks are a little puzzling. At the removal of this tax it is not recorded if Edward then set about obtaining a van with road springs for the better comfort of his passengers — in keeping with private carriages and stage coaches of the period.

We have seen how Edward moved to Feering early in the century and started his business in 1815. In the following year, 1816, he became licensee of The Bell Inn — a position held until 1837. A picture of the inn taken some time later shows some outbuildings alongside an open space at the side of the inn, see Pic.2. These could well have been stables and a place to park his cart, as well as a storage area for goods prior to delivery or collection. Land at the rear of the inn may well have provided grazing for his horse. Taking up the licence of the inn not only provided an additional income but also, as a tenant of the property, a home for Edward and his family. Becoming a licensee was a shrewd move since it brought him into contact with business people and the farming community. To entrust the carriage of their goods to someone they knew and could rely on brought satisfaction. This led to Edward gaining even more confidence and achieving a reputation for reliability which was passed down to later generations running the business.

When Edward Moore became landlord The Bell Inn was owned by Isaac Brightwen, a brewer from Great Coggeshall, together with his partners Robert Tabor, who was a brewer from Colchester, and Bartholomew Brown, a Colchester merchant. In 1828 the Brightwens were declared bankrupt and The Bell Inn was sold at auction in The Chapel Inn, Great Coggeshall. Today such uncertainty has been removed, and not only does The Bell Inn survive but is thriving under the ownership of Gray and Sons, the brewers, as it has been for the past 70 years.

However, in 1828 things were very different and the upheavals of the time must have been very unsettling for Edward and his family, but he weathered the storm and continued as the licensee with the new owners. The fact that Edward ran The Bell Inn for 21 years is proof that he was successful and was a demonstration of his business acumen. Clearly he was a popular and astute publican which must have been of great benefit to his carriers business. During this time the business was growing for we know that by 1825 his cart was running twice weekly from Feering to Colchester and to Chelmsford through Kelvedon.

The days of operating were set to coincide with market days in each of these towns. Departure and arrival times were to a fixed timetable. For Colchester, the cart would leave Feering at 8.00 a.m. on Wednesdays and Saturdays, arriving at 11.00 a.m. at The Horse and Groom in Crouch Street and then leaving at 6.00 pm for its return journey to Feering. Sadly The Horse and Groom was demolished with the widening of Balkerne Lane in 1959. For Chelmsford, it would travel on Tuesdays and Fridays leaving at 8.00 a.m. and arriving at 12 noon at The King's Head then leave at 6.00 p.m. for Feering. The site of The King's Head is now occupied by Woolworths in the High Street. When The King's Head was pulled down in 1931 the carrier — by now a motor van — terminated at The Two Brewers in Springfield Road, a public house which is still on the same site today.

The importance of the local carrier to the community should not be underestimated. Here was a servant who was known locally, who could be trusted and was reliable. The attraction of the carrier visiting towns on market days was that villagers had a source for obtaining household goods cheaper than they could buy from the local shop. Collecting shopping for those unable to travel with the carrier and having goods delivered on the same day was an asset and welcomed by many. This service continued well into the motorised bus era as many readers may well remember. Although the cart did not generally detour from the route it only needed someone to be waiting by the roadside to pick up the items ordered. The time when the cart would arrive could be predicted within reasonable limits. With a maximum permitted speed of 4 m.p.h. when carrying passengers the journey from Colchester to Feering would take about 3 hours. We know from a later interview with the family that Edward was caught on three occasions exceeding the speed limit and fined £7 each time.

The success of Edward choosing market days for the service can be gauged by rivals appearing on the same days and often leaving at the same times. Undoubtedly Edward Moore was facing increasing competition, with the Chelmsford run being very intense with as many as six other operators at one time. However, some of these may have been waggons. One later entrant, who was a very persistent competitor, was Higgleton, who operated the same routes from his house in Kelvedon. Over the years different members of this family owned and operated the cart but by 1870 they had given up. Others appeared and disappeared and by the turn of the century only the Moore family had survived.

As we have seen Edward gave up running The Bell Inn in 1837 and finally decided to retire in 1844 when he handed over the carriers business on 20th February of that year to his son, William Edward, who had already started to fill an important role in the business.

Edward spent many years of contented retirement at his home in Little London, Feering, and died on 6th June 1873. He is buried in Feering All Saints churchyard, as is his wife, Mary Ann, who died in September 1843. By living to the grand old age of 85 years he had not only witnessed the handover of the enterprise to his son, but also to his grandsons, Charles and Edward, who brought prosperity to the business in the latter part of the century. Certainly this must have been a triumph for his pioneering spirit.

3 An actual carriers cart. Mr Clapson of Great Bardfield, Essex.

Chapter 2

Consolidation

4 The George, High Street, Kelvedon.

We have seen in Chapter 1 how the founder relinquished his position as licensee of The Bell Inn in 1837 but continued to run the carriers business until his retirement in 1844 when the business was handed over to his son, William Edward. Operations remained in Feering until about 1844 when they were then transferred to Kelvedon. With William Edward already living in Kelvedon, this was a natural step and had the advantage of bringing the business nearer to the railway station which had opened in Kelvedon in 1843. About this time William Edward's home was just a few doors down from The George in the High Street. Nearby was a small yard where, in all probability, the horse and donkeys were stabled and the carriers cart housed.

Having taken up the reins it was now up to William Edward to ensure that the gains made by his father were firmly established. But before we continue with the story of the carriers business we first need to look at William Edward's family life up to 1844. We have seen how William Edward was born just a few weeks after his father had started the carriers business. Reference to the population census of 1831 shows William Edward listed as a carrier and so we can presume he joined his father's business on leaving school. With good fatherly attention he would have been well placed to take over the business in 1844 when he was 29 years old.

Meanwhile, in April 1837, William Edward married Rebecca Osborn. It is interesting to note that the Moore family were members of the Church of England until Rebecca joined the Congregational Church in Kelvedon on 30th March 1843. William Edward was later converted to this belief and all subsequent generations of the family with which we are concerned followed in their footsteps. Later the Church became known as the United Reformed Church. If there were any doubts about William Edward's ability to carry the business forward, there were none as far as fathering children was concerned since in his 29 year marriage he produced no less than 18 children. Rebecca, his wife, died in January 1856 leaving eight children, as reference to the epitaph on her tombstone will testify.

In November 1856 William Edward married a Mrs Sarah A. Smith who originated from Shoreditch, London, where they married in the local Church of St. Leon. The significance of this marriage is that Sarah (Ann) continued the business after William Edward's death in 1865 before his sons Charles and Edward took over in 1867. The reason for the delay was that William Edward had directed that his estate was to be put in trust for the benefit of his wife until her death when all would be inherited by his children. The estate included the carriers business and the rent from four cottages in Tolleshunt Knights.

At the time of William Edward's death his two elder sons, Charles and Edward, were already working in the carriers business and therefore it was inevitable they would take over the running of the business on a day to day basis. Clearly an arrangement was needed with the Trustees which would recognise this reality. Undoubtedly this accounted for the delay of two years before the business was in the name of Charles and Edward, who took over in 1867. Sarah, their step-mother, died on 21st July 1882.

If we now return to the business under William Edward, we would expect to see a growth in trade due to the arrival of the railway in Kelvedon in 1843. For although the stage and post coaches still brought businessmen to the area, the speed and comfort of the railway made Kelvedon an attractive terminus for surrounding places such as Coggeshall, Tiptree and Tollesbury. Apart from an increase in passenger traffic, the volume of goods would have increased also. The increased speeds and reliability of the railway meant stocking levels held in local shops could be reduced, giving rise to greater frequencies of distribution. It would seem William Edward did not, or could not, seize these opportunities for expansion. It would take the next generation of Charles and Edward to achieve this.

Even so, William Edward achieved a level of respectability in business circles, and during his stewardship we find a van had replaced the carriers cart, and donkeys arrived to supplement the one horse. Much has been written about these donkeys being used in a unicorn form (one donkey in front of the other two) to haul the van to Chelmsford. The unicorn form was unusual. Normally, when horses were used and more than one was needed, they would be harnessed either as a pair, or four in hand. It is said that donkeys were used to avoid payment for a posthorse licence, but we need to look at these comments a little more carefully. Excise duties on horses commenced in 1784 and over the years the rates of duties varied. In 1853 the rate for a horse exceeding 13 hands was £1.15s.0d., whereas a horse or mule not exceeding 13 hands was 10s.6d. It can only be assumed donkeys were either exempt from this duty, or the lower rate would have applied.

Duty would also have been payable for the carriers van in the form of a Carriage Licence and from 1853 this amounted to £2.6s.8d. At the time of William Edward's death in 1865 the firm only possessed one van, one horse and three donkeys. With both the Post Horse and Carriage duties being repealed in 1869, this was an added incentive for expansion which Charles and Edward carried out with great zest when they became Jobmasters as well as Carriers.

Chapter 3

Time to Expand

We have seen how Charles and Edward's father died in 1865 and although the brothers were already in the carriers business they were prevented from taking over due to the terms of trusteeship set up by their father shortly before his death. Whether they were in serious conflict with their stepmother is not known, but it took two years for them to gain effective control of the business. The delay in takeover until 1867 was unfortunate because it later became quite obvious the entry of these two brothers, the third generation of the family, was to have a profound effect on the fortunes of the carriers business.

Earlier on we saw how in 1825 the carriers cart was operating twice weekly on market days to Chelmsford and Colchester. In 1870 this timetable was still being maintained, but by 1878 the frequency had increased to three times a week to Colchester and five times a week to Chelmsford. The carriers cart now reached Colchester on Mondays as well as Wednesdays and Saturdays, and Chelmsford on Mondays, Tuesdays, Thursdays, Fridays and Saturdays — Chelmsford

5 Carriers cart and donkeys, Swan Street, 1935. Scene must have been reminiscent of similar occasions almost 100 years earlier.

obviously generated more business. Growth was now very rapid because we find by 1876 a new stable with a loft over it had been erected, with the horse-drawn fleet comprising: a Brougham, Phaeton, Cab, Van, Cart and five horses. The house in Swan Street in which Charles and his wife lived was sufficiently large to bring up their family, but equally important was the yard behind the house which was able to accommodate the enlarged horse-drawn fleet. These were the business premises of C. and E. Moore with their name over the doorway leading to the yard. By now the brothers, Charles and Edward, had clearly entered the private hire business and were making full use of hiring out horses and carriages as well as taking the extra trade available from the railways. The firm could rightly claim they were now 'Jobmasters and Carriers'.

As part of this enterprise a contract was entered into for the provision of horses and a driver for the newly arrived Kelvedon fire engine in 1897.

6 Brand new Fire Engine 1897. William Moore in drivers seat. George Braddy, Fire Captain, central figure standing.

KELVEDON

FIRE ENGINE

A MEETING

WILL BE HELD IN THE

PUBLIC HALL,

On Wedneday next, May 25th,

For the purpose of receiving the Report of the Committee, and for arranging for the working of the Engine.

The Chair will be taken at ½-past 7 o'clock, by

MR. GALPIN.

W. GODFREY, PRINTER AND BINDER, KELVEDON.

Fig. 1 A poster from 1897.

A picture of this fire engine with William Moore in the driving seat is shown in Pic. 6. As we will find later, William was the son of Charles and one of three brothers to take over the carriers business in 1905. These brothers were to continue the contract with the Kelvedon Fire Brigade until the arrival of the Ford motor fire engine and pump circa 1930. From the picture it can be seen that the fire engine was in fact a horse-drawn manual fire pump, which cost £91. 0s. 0d., and was purchased from the makers, Shand, Mason and Company. Funding of the appliance was raised through local subscription following an outcry after Mr Wiseman's Kelvedon School was burnt down, when it was believed to be quite disgraceful that a fire engine had to travel from Colchester to attend a fire in Kelvedon. Mr Wiseman's school was situated opposite The Angel Inn and where the Convent building now stands. Some years later Jimmy Pitchford became a driver of the horse-drawn fire engine and gave an account at his retirement of a hair-raising experience of reaching a fire at Marks Tey within 17 minutes of receiving the telegraph message. As the distance covered was over 4 miles, which approximates to 16mph, it was hardly surprising that the Fire Captain vowed he would never ride behind that driver again! However, in spite of this threat he did so many times.

It is perhaps worth noting that Jobmasters who were in the business of hiring out horses for stage and mail coaches satisfied a very different need. This was part of big business which grew out of increased speeds of coaches, necessitating frequent changes of horses. Generally such business was conducted through long term contracts between operators and owners of post-houses and inns. The Angel, The Star and Fleece and

The Swan, all of which were in Kelvedon, were typical of those providing such facilities, as well as being hostelries. The Moore brothers were not part of this enterprise of course, but served a smaller market providing similar facilities for the private traveller. The abolition of the post horse license in 1869 helped to reduce costs, thereby increasing competitiveness and stimulating business growth.

In 1879 Charles and Edward took their boldest step yet when they built the 'Moore's Temperance Hotel' in the High Street, which most people will remember as 'Roslyn Garage' — the name given to the site after the arrival of motor vehicles. The building of this hotel provided accommodation for the many traders arriving by the railway, and complimented the carriage business already established for the conveyance of travellers to local areas. The premises with stabling for 20 horses was an essential prerequisite for the expansion of the post horse business. The stables were not only needed for the firm's own horses, but also for travellers requiring liveried facilities as part of their journey. A picture of the hotel taken in 1905 appears in Pic. 7. On the side of the building is a sign which states 'good stabling' — the aim of this being to give confidence to the travellers that their horses would be well cared for whilst in these liveried stables. The two boys in the picture, Ernie Dale and Jimmy Pitchford, were stable-lads who went on to complete 50 years of service before retirement. Part of their duties at that time would have been to ride postillion on Moore's horses whilst the coachman continued to ride on his own carriage. The carriage house gave space for a number of carriages of varying types some of which were listed earlier in this chapter. Pictures of a horse-drawn hearse and other carriages are shown in Fig. 3.

The land on which the hotel, carriage house and other buildings were erected, originally belonged to the property known as 'Strogulls' — part of which became known as 'The Grotto'. By the time the land was purchased by Edward and Charles in 1879 it had become the garden of The Star and Fleece. A further acre of land behind the garden and extending beyond the nearby Friends Meeting House was also acquired about that time. Another plot of land was added in 1887 and a site plan drawn up in 1940 shows the extent of the premises through all of these purchases. A copy of this plan is shown in Fig. 4. The boundaries were to be further extended during World War II when the garage entrance was widened and the Friends Meeting House with adjoining land was purchased.

7 Moore's Temperance Hotel c1905. Stable lads Ernie Dale (left) and Jimmy Pitchford.

KELVEDON, ESSEX.

Temperance & Commercial Hotel,

(WITHIN THREE MINUTES WALK OF THE STATION).

Gentlemen and Visitors will find every Comfort at Moderate Charges.

C. & E. MOORE, Proprietors.

C. & E. MOORE, Jobmasters and Carriers. Open and Closed Carriages on Hire. Omnibus for Private Parties. Good Stabling. Orders by Post or Wire will receive prompt attention. Our Carriages meet all Trains.

PROPRIETORS OF THE COGGESHALL OMNIBUS.

For Times see G.E.R. Time Table.

Fig. 2 Extract from 'Harrod's Special Directory of Great Britain', 1886.

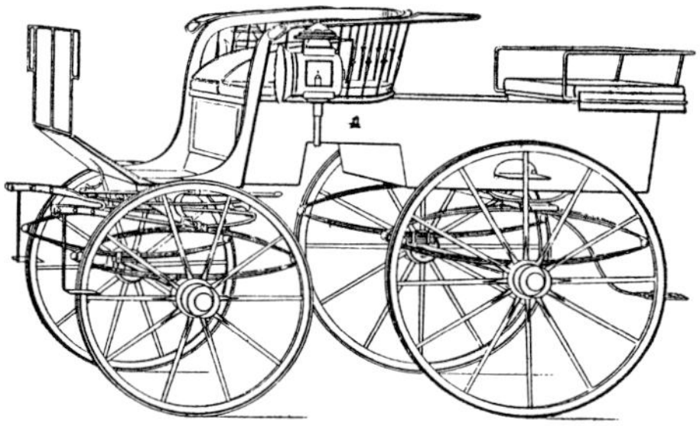

PHAETON

VICTORIA

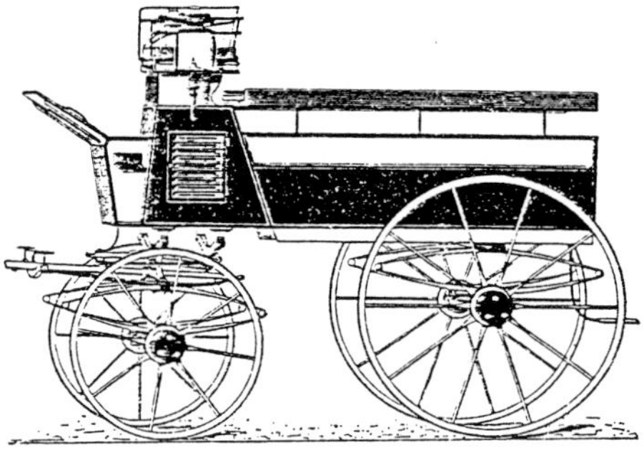

WAGGONETTE

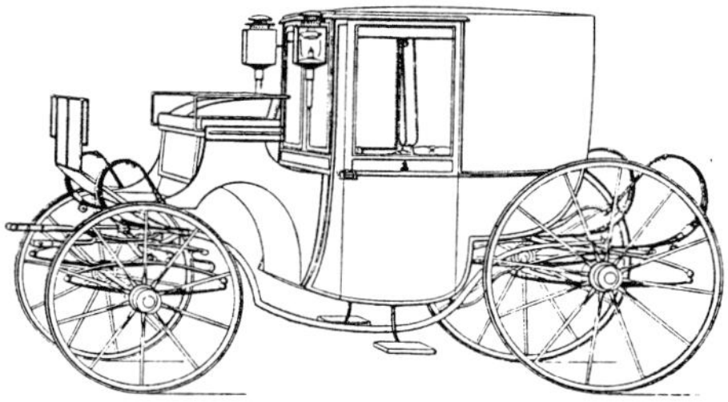

BROUGHAM

HEARSE

Fig. 3 Horse-drawn carriages and hearse of the period.

Fig. 4 Site plan of garage and other property owned. 1940.

The brother's crowning event however was the introduction of the horse-drawn omnibus service in November 1881 between Kelvedon and Coggeshall. The significance of the service is that it led directly to the build-up of a network of stage carriage services which became the hallmark of Moore's success. However, the service initially faced great difficulty through the Great Eastern Railway refusing to allow the bus to ply for hire on railway property. Such an attitude was unheard of to Charles and Edward who sought to solve problems through friendly negotiation and not by employing strong-arm tactics. Solicitors from

Coggeshall were engaged to put their case to the railway board, following which Charles Moore was invited to a meeting with the GER directors which included the Chairman, Mr C. H. Parkes. One of the railway directors stated their objection to the bus was that by being on the high road it was in opposition to the railway. But Charles Moore said he could not accept this since the road van had been running before the railway, and therefore it was the railway who were in opposition. Mr Parkes then proposed that the bus could go to the station upon payment of a 10s. 0d. rent, but the other directors would not agree to this.

The whole story was related in a press interview given by Charles Moore which appeared in the Essex Herald of 2nd March 1915. The events which followed the meeting with the railway board then moved to Kelvedon and because they are particularly interesting an extract from this part of the interview is given below;

> "........ They (the railway board) then sent two inspectors down to investigate the case, and these inspectors hired our own cab from Kelvedon to Coggeshall and rode back by our 'bus! My brother was the conductor of that day and I the driver. They had a conversation with my brother and asked him what we carried. He told them anything from a barrel, live calves, and quarters of beef to the ordinary parcels. Said one of the inspectors, 'Dear me. I will tell you what we will do. We will give you the carting from Kelvedon station to Coggeshall, the delivery in Kelvedon, and the shunting if you will give up your road.' My brother replied he was sure I would not give up a certainty for an uncertainty. They went back and the Company issued tickets to Coggeshall, chartering a 'bus in opposition to ours, but that did not keep on more than six months, although their fare was 6d. inside or out, and ours was 9d. inside and 6d. outside. When their bus fell off I wrote to the Company asking if they would admit us to the station and put our time-table in their time-table, it having been proved that we had raised the railway traffic? Almost by return of post came a reply that they would do as we asked, and ever since our time-table has appeared in the railway time-table."

LIST OF COACHES

Plying between Neighbouring Towns and the several Stations. Given for the information of Passengers, but for the accuracy of which the Company is not responsible.

From the undermentioned Stations	Distance from Station	To	On arrival of Trains leaving Liverpool Street at	From	At	In connection with Trains due in Liverpool Street at
	Miles					
Kelvedon	3	Coggeshall . . .	9.3 a.m. and 3.35 p.m., on Week Days.	Coggeshall . . .	8.30 a.m., 12.10 p.m., and 6.40 p.m., on Week Days.	10.10 a.m., 2.10 and 8.53 p.m.

Fig. 5 Extract from GER timetable November 1883.

The Kelvedon omnibus appeared in the GER timetable for the first time in November 1883 and an extract showing this is given in Fig. 5, which also shows the Liverpool Street station train arrival and departure times.

It is possible, of course, that the railway set out to protect their interests, and later on the local press reported rumours that the GER were starting motor bus services in the surrounding areas of Colchester, Chelmsford and Clacton-on-Sea. Eventually these services did come into operation but not until 1905/1906. So after six months of wrangling the bus was allowed to go to the station, but it was eleven years before cabs were permitted to ply for hire on railway property. This account illustrates the railway board's sensitivity to competition from the road which continued well into the era of the London North Eastern Railway, as typified by the instance covered in Chapter 7. There were many such instances of objections by the railway companies during the period of bus regulation as Moore Bros. was to discover with some of their licence applications.

The Moore brothers owned three single-deck horse-drawn omnibuses, the first of which was acquired from a Rochford owner after the service between Rochford and Chelmsford had been given up following the terrible blizzard in January 1881, when many lives were lost. The bus was bought for £5.0s. 0d. and the horse for £3. 10s. 0d. Not all three buses were of the same size — the two smaller ones seated 6 - 8 passengers inside with 2 riding outside, whereas the larger bus carried 8 - 10 passengers inside, and 5 outside on two rows of seats, one of which was partly taken up by the driver. The smaller buses were pulled by a single horse, with the larger bus being hauled by a pair of horses. Eddie Warner who worked for the company for many years served as a conductor on the Coggeshall bus and well remembers those early days. The late Doris Moore, who sadly died in January 1998, happily recalled travelling between Coggeshall and Kelvedon to attend dances with her future husband, the late Eric Moore. In fine weather she preferred to travel outside along with the driver. Regrettably no pictures of these buses appear to have survived but an example of a similar type is shown in Pic. 8. The horse-drawn buses were all withdrawn by 1924.

TEMPERANCE HOTEL
KELVEDON ESSEX

G. & E. MOORE,
Jobmasters & Carriers,
ESTABLISHED 1815.

OPEN AND CLOSED CARRIAGES ON HIRE
omnibus for private parties
GOOD STABLING

☞ *Our* CABS *are not allowed at Kelvedon Station* UNLESS ORDERED
[over]

September, 1884.

OMNIBUS TIMES.

LEAVE KELVEDON RAILWAY STATION—	LEAVE MARKET HILL, COGGESHALL—
7.45 a.m.	8.30 a.m.
10.25 a.m.	12.10 p.m.
5.20 p.m.	6.40 p.m.

Fares—Single Journey 6d. Inside or Out.

Also see Times G.E.R. Time Table.

Fig. 6 The first known omnibus timetable, 1884.

8 Typical station omnibus having a pair of horses and two rows of seats on the outside.

Chapter 4

100 Years On

8a Carriers van and pair. Driver Horace Moore. Passenger Alice Cobbold. c1900.

Following a succession of achievements the brothers Charles and Edward decided at the ages of 64 and 61 respectively to hang up their hats and retire. They had every reason to go out with a fanfare of triumph. On 24th March 1905 Charles Moore gave a supper to all employees of the firm to commemorate the recent marriage of his daughter, Anne. However, at the end of the evening Charles made an announcement that he and his brother Edward were retiring the following day. Thus on 25th March an Indenture was drawn up conveying to Charles' three sons, William Charles, Horace Bramley and Basil Wilberforce '.....All that tenement used as a dwelling house and Temperance Hotel with the yards, stables, coach houses, and other outhouses and buildings, together with that parcel of land containing one acre, one rood and thirty eight perches at the rear of the property described above.....'

For this, the retiring brothers were to be paid £500 each, making the value of the business at that time worth £1000, approximating to a current purchasing power of £57,000. Charles' sons had inherited well, but at the same time they must have felt a sense of duty and responsibility to carry the business forward to greater success. They had to look no further than their father and uncle for inspiration. We have to remember that in 1905 everything was still horse-drawn and indeed use of horses did not entirely disappear until very much later. Even so, the brothers had arrived in an age of technological change and would have to adapt accordingly.

9 Kelvedon High Street, 1870. Showing pole carrying telegraph wires behind Star and Fleece sign. Telegraph was used until arrival of the telephone in 1908.

Firstly, the Kelvedon telephone exchange opened on 1st March 1908 when it became a sub-exchange of Witham. It operated only between 8.00 am and 10.00 pm. Moore's acquired the number 6 and so were one of the first to be connected — the Post Office being number 1. Perhaps somewhat surprisingly availability of this means of communication was not more widely advertised by the Moore brothers. As late as 1915 the telephone number was still not appearing on their business card, nor on

10 Typical rural Royal Mail cart.

the timetables of that period. Prior to the telephone the only means of wire messages was through the telegraph system where the message had to be handed to the local Post Office for onward transmission. A pole carrying telegraph wires can just be seen behind the very elaborate Star and Fleece Hotel sign in Pic. 9. The picture was taken about 1870, before The Temperance Hotel was built.

Secondly, the brothers had witnessed the dawn of the motor car and must have often wondered how long it would be before they had to adopt this form of passenger carrying. In some respects it posed an attraction, for horses were very expensive to keep but on the other hand motor vehicles were unreliable at that time and breakdowns frequently occurred. The first car to arrive was in 1912, although the horse was still in great demand. A contract was won to supply a driver and horse to haul the Post Office cart to deliver and collect mail between Kelvedon and Coggeshall, which also involved bringing and collecting mail from the trains. Eddie Warner, who joined Moore's when he was only thirteen years old, drove the cart for some while having taken over from George Buck senior. A typical rural Post Office cart of the period is shown in Pic. 10.

Horses and carriages were also supplied to undertakers in the area including Birkin's of Coggeshall and George Braddy of High Street, Kelvedon. Initially a waggonette was used to carry the coffin but by 1905 it was replaced by a 'Shillibeer'. George Shillibeer is best known for operating the first omnibuses on the streets of London in 1829. Later he became an undertaker but he was also a coachbuilder and in 1842 his

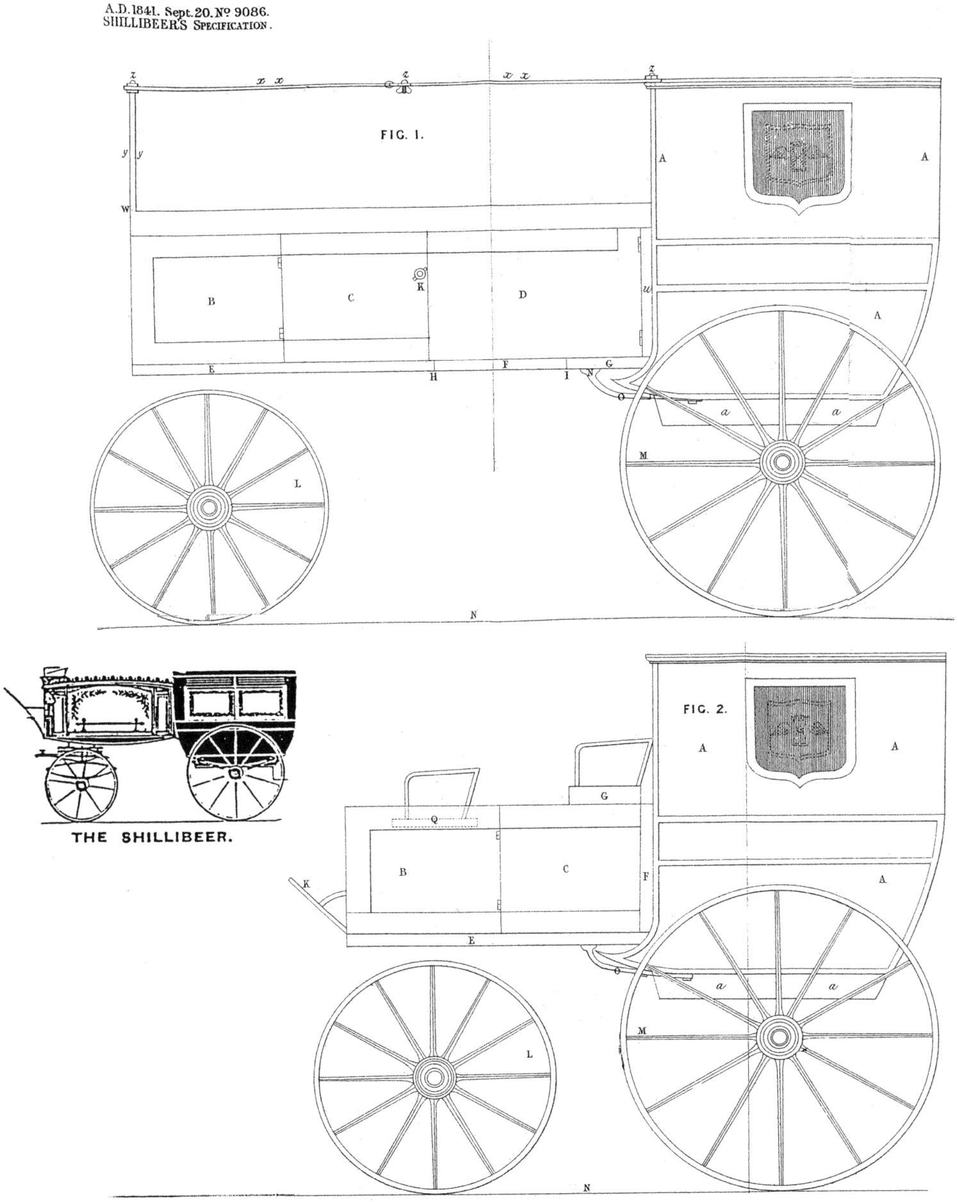

Fig. 7 Shillibeer combined hearse and mourning coach. Insert shows later design.

invention of combining a hearse and mourning coach was patented. The aim of the design was to reduce funeral costs as only one coachman and one team of horses would be needed instead of two. The idea was quite ingenious in that following internment, the undercarriage could be foreshortened through a winch operating a long screw, thereby freeing the box on which the coffin was carried, and so allowing it to be dismantled and turned into seats for the bearers on the return journey. This was not a great success and a later model resembled a Victorian glass-panelled type of hearse. Part of the drawing which accompanied George Shillibeer's patent application, with an insert showing the later variation, is shown in Fig. 7. About 1915 a motor hearse was available, but for those who continued to favour the use of a horse, a traditional Victorian hearse was retained, see Fig. 3 in Chapter 3. The second motor hearse was a new Model T Ford chassis delivered in 1925 with a custom built body. This survived until about 1935 when a second-hand Chrysler arrived, which in turn was replaced with a new very smart looking Humber Pullman in 1950. The association with George Braddy continued well into the 1950's.

The carriage house and stables behind The Temperance Hotel were built to accommodate 20 horses and 20 carriages. At the time the buildings were constructed in 1879 no one could ever have imagined that they would ultimately be given over to the garaging and servicing of motor vehicles. As we have seen the first car arrived in 1912, which was a Model T Ford painted blue. However, before we enter the motor era perhaps we should pause for a moment in our story and look at the family history from William Edward's time, whom you will remember was the founder's son. From a family of 18 children only Eliza, Charles, Edward and Robert made it into adulthood. Eliza was born in 1839 and died in 1924. Charles and Edward were born in 1842 and 1845 respectively and were the only two brothers to enter and carry on the business, as we have already seen in Chapter 3. In 1863 Charles married Susannah Ketley — they had six children, of whom only William, Horace and Basil became part of the firm. Edward was married in 1869 to Caroline Roofe, and there were two children, neither of whom joined the business.

With William, Horace and Basil having bought the carriers business in 1905, it is important we look at their family backgrounds because we have now arrived at a point in our history where the children of William and Horace are about to join the business. This will be the fifth generation and the last to own and run the company.

William's eldest son, Eric Charles William, was born in 1897 and after leaving school trained in London as an accountant, but with the advent of the First World War he joined the Navy and saw service until almost the end of hostilities. He then joined his father's bus company and we find that on 2nd July 1918 the Borough of Colchester issued him with a conductors licence. Over the next few months we then find Eric employed as a conductor on open-top double-deckers between Kelvedon and Colchester. However, he did not settle and by one of those twists of irony his fortunes were shortly to change.

Round about 1920 the washerwoman of The Temperance Hotel was having great difficulty in getting the household linen dry due to poor weather and persistent rain. As she was complaining Eric's father happened to be in hearing range of these moans and groans and declared, 'What we need is a duddy good laundry round here'. 'duddy' was a word unique to William and the nearest he ever came to swearing! At this time Eric was courting Doris Croyden whose parents owned a general store on Market Hill in Coggeshall. Doris' mother had already purchased the school in West Street, Coggeshall, which later was to become the premises of the laundry. To finance the laundry £100 each was contributed by Eric Moore, his mother Rosanna Moore, Doris Croyden and Mrs Croyden. Thus the laundry was started with £400, which currently approximates to a purchasing power of £8,200. As they say the rest is history!

Today Eastern Counties Laundries Ltd. is a multi-million company with factories in Clacton and Chelmsford. The main thrust of the business is the provision of linen to the major hotels in London and elsewhere under long term contracts. I know that Dudley Moore, who is the present managing director and son of Eric Moore, feels an affinity with the bus company whose finances kick-started the laundry to the great success enjoyed today. And so whilst the bus company took a financial interest in the laundry business, by the same token the laundry was later to gain financial interests in the bus company. However, this was to end in 1940 as we shall see in Chapter 9.

Eric's brother, Harold Ketley, was born in 1901 and had visions of becoming a draughtsman on leaving high school, but these ideas soon changed when he joined the bus business. We find in the Colchester Borough licensing records that a drivers licence was granted to him on 6th July 1921, which was in keeping with tradition where his ancestors, as well as running the business, had served time driving and conducting on the horse-drawn and motor buses. His change of heart was fortuitous for the firm since his clear and analytical mind was put to good use with the arrival of bus regulation in 1931 at a time when considerable bureaucracy had to be faced. In the 1930's he was also responsible for bus scheduling and compiling published timetables. On the death of his father in 1940 Harold became a Director and General Secretary of the Company — a position held until closure in 1963.

Percy Horace, the son of Horace, was born in 1898 and served an apprenticeship with Hunwick Brothers, who were Engineers and Brass Founders in Station Road, Kelvedon. At the outbreak of The Great War in 1914 he joined the armed services and was posted to the Motor Transport Depot at Upper Norwood, London. On cessation of hostilities he completed his apprenticeship and then joined the bus business as an engineer, subsequently becoming head of engineering and in 1940 a Director of the Company, in which capacity he continued to serve until 1963. It is interesting that Clayton Hunwick, who was a steam engineer, was the very first driver of the steam bus when it was delivered to Moore's from Clarkson. Horace Moore had asked Clayton to drive the bus to Colchester and back to ensure 'all was well'.

During the period 1912 to 1919 the Model T Ford became the mainstay of the Moore's fleet until the arrival of the first Thornycroft double-decker in 1919. The Model T Ford which was very popular and found favour with many operators, was cheap, fairly reliable and economic. As mentioned earlier, the first Model T Ford arrived in 1912 and was registered on 17th September with number F6060. It carried a 'little laudaulette' body which was painted blue, and some five years later we find it still in Moore's ownership, but now painted black with biscuit coloured wheels. Indeed this was the car which

11 A young Ernie Dale with a Moore's horse.

gave the owners and staff their first attempt at handling a motor vehicle. For some this was an ordeal, and in the case of Basil Moore he never came to terms with these 'modern contraptions', and much preferred the placid temperament of the horse. For this reason, and perhaps others, he was content to continue with the carriers side of the business.

Undoubtedly some employees felt their jobs would never be quite the same without that special working relationship between man and horse. Reference to Pic. 11 shows Ernie Dale as a young stable boy with a Moore's horse and this conveys the meaning very well. Sadly this was lost for ever when Ernie became an early bus driver. William and Horace must also, at first, have had some misgivings about the new mode of transport since the Model T Ford was acquired originally on a trial basis only. However, commercial considerations obviously dictated otherwise and the way forward would be with motor vehicles. Nevertheless, customer's needs could not be ignored for there were some who would not be persuaded to ride in these modern miracles of science. This alone would prolong the transition period and for some time to come both forms would be needed. Later in 1912 it is said the first bus was purchased, which was a 14 seater chain-driven Daimler. It was a second-hand vehicle and obtained from the father of the circus magnate Bertram Mills. Regrettably all attempts to trace this bus have so far failed, but research

12 Frank Tebbell at the wheel of his ex-Moore's Daimler.

13 1950's picture of Frank Tebbell (with drivers cap) then with Quest of Maldon.

continues. It is now known that the father of Bertram Mills was Halford L. Mills who owned a carriage building business in Paddington, London. Therefore it is possible the bus carried a body from that firm.

In 1912 the only published stage service operated by Moore's was the daily horse-drawn bus to Coggeshall. As we know this was inaugurated in 1881 and terminated at the Woolpack Inn. Therefore, initially the ex-Mills family Daimler must have been used only for private hire work. It was to be another ten years before the next Daimler arrived which was a CB type chassis and carried a 36 seater double-decker body. Undoubtedly this was an ex-war department chassis since production of this type ceased in 1917. The London registration, XL5891, was taken out in 1922 and therefore it can be reasonably assumed Moore's were the first owners in its civilian life. The bus remained in the fleet until about 1926 when it was sold to Frank Tebbell of Maldon who converted it to a charabanc and had pneumatic tyres fitted. At the time Frank was working full time at Crittall's, Witham, and being full of enterprise used the charabanc to convey fellow workers from Maldon to the factory, for which he made a charge of 3s. 6d. for a weekly return ticket! Otherwise it was used for private outings and Frank Tebbell's daughter, Mrs Barbara Whittaker, can well remember travelling on the bus with her father to Clacton and Southend. She also recalls the leather seating getting very hot in summer and all passengers hastily pulling the hood over them when it started to rain. One such outing is shown in Pic. 12 where the party are waiting to move off from outside the Heybridge Headquarters building, which was sited on the Maldon side of the Wave Bridge.

The ex-Moore's Daimler was the only passenger vehicle owned by Frank Tebbell, who garaged it in a building at the rear of The Welcome Sailor public house at the bottom of Market Hill, Maldon. What finally happened to the Daimler is not recorded, but we know that later on Frank Tebbell joined Quest Motors of Maldon as a coach driver and often drove their AEC Regal on the Maldon to London Kings Cross service. Quest Motors was another independent bus operator whose proprietor was J. W. Gozzett. The passenger carrying side of the business was sold to Eastern National in 1934, but the Gozzett family continued with the private car hire business. After World War II they were again using coaches under the fleet name 'Quest', but with operations now limited to excursions and tours, and Frank Tebbell became their senior driver. A picture of coaches waiting to depart from Heybridge during the 1950's is shown in Pic. 13.

As we will see in the following chapter the first regular motor service was operated by the steam bus between Kelvedon and Colchester. This alternated between two routes via Tiptree and Coggeshall. It should be noted that the carriers van was still running along the main road to Colchester but now three times weekly. On 10th September 1919 a service via Coggeshall, Earls Colne and Halstead was started, using open-top double-deckers, but on Wednesdays only and just one round trip, arriving Braintree 11.15 am and departing 3.30 pm. Later on a shortened run was added between Halstead and Braintree, leaving Halstead at 12.30 pm and returning at 1.15 pm. Investing in larger capacity buses was paying off and so further Thornycroft open-top double-deckers were bought in. Apart from opening up new routes, the frequency of existing ones was also increasing. In 1919 we find the Colchester service via Tiptree operating each weekday apart from Thursday (early closing) — the Coggeshall service running two days a week remained unchanged, presumably due to competition from Arthur Hutley's bus which operated from Coggeshall. The carriers van to Chelmsford was operating five days a week, as it had been from about 1870.

Obviously the expansion of the bus fleet was very necessary to support these new demands. The Thornycroft double-decker already mentioned was open-topped and purchased second hand from A.W.Berry of Colchester — an operator well known in the Colchester area who started his business on Christmas Eve 1904 by running a bus to Mersea. The speed with which this bus was put into operation by Moore's was quite remarkable and demonstrated the urgency of additional capacity.

14 Thornycroft at Market Hill, Coggeshall, May 1919.

Delivery took place on 8th May 1919, the name 'Berry' being painted out the following day and 'Moores' inserted, with 'Phone 6' being squeezed into the space remaining. The words 'Colchester and District' stayed on! The bus was used almost immediately on private outings and a picture of such an outing from Coggeshall Market Hill is shown in Pic. 14. The fact that the bus was around in Berry's colour of brown for a while was not too out of place since Moore's colour at that time was biscuit. However the bus was completely repainted later on.

In spite of motor bus and horse-drawn carrier expansion the private hire side of the business was far from being neglected. With speeds now allowed of 12 mph more distant places were achievable for a days outing, and attention would have to be given to meet this increased demand. Of course, the horse-drawn bus — or the omnibus as it was referred to in timetables and other publicity material — continued to be available for private hire. Naturally, one of these buses would always be required during the week for the service between Kelvedon and Coggeshall, as described in Chapter 3. The original Daimler, referred to earlier, gave a

15 Thornycroft charabanc with Holmes and Smith, Southend, before sale to Moore's in 1915.

seating capacity of 14, but then a Thornycroft charabanc was purchased in 1915 which provided a capacity of 20 seats. This vehicle was purchased second-hand from Holmes and Smith Ltd. of Southend-on-Sea, who were very well known operators of tours and excursions, and used the fleet name 'Royal Red'. Their charabanc fleet comprised a number of Thornycrofts and a picture of the Moore's Thornycroft whilst still with Holmes and Smith can be seen in Pic. 15. Holmes and Smith Ltd. was the forerunner of Westcliffe-on-Sea Motor Services Ltd. Charabancs were very popular both at this time and well into the 1920's and so the Moore brothers could be well satisfied with their progress and they would end the era in good shape. Gradual expansion would bring success but competition was increasing and this was yet another challenge that would have to be faced during the next decade.

For Basil Moore the development of the passenger carrying side of the business clearly was not to his liking. The origins of the business lay with the carrier and it was this aspect of the business which appeared to interest him most. Clearly with a conflict of interests it was decided that his connection with the Moore's bus firm should be severed. And so on 8th May 1919 an Indenture was signed dissolving the interest which he had inherited from his father and uncle in 1905. In compensation he was paid £450 — which approximates to a current purchasing power of £9,300.

Following this Basil formed a partnership with Frank Doughton who already operated a haulage business from Swan Street, Kelvedon. The carriers business traded under the name of 'Moore and Doughton'. Their first vehicle was a Model T Ford lorry, which was purchased second-hand in June 1919. Originally this vehicle, F4806, had been a 2 seater car and was new in 1911. It was then owned by John René Courtauld of Braintree. By 1916 it had been sold to Bertrand William Deal of Brooklands, Kelvedon, who arranged for it to be converted to a delivery van and painted green. Later on a Garford lorry was purchased. By 1926 the partnership had been dissolved and Basil continued the carriers business under his own name of Basil W. Moore. It is interesting that in September 1927 he purchased a new Morris van, VW1290, which was registered 'hackney and goods' and had 14 seats. Much later still, he replaced this with a similar vehicle which had been with the ambulance service in Witham — presumably for reasons of its low mileage. He could now rightly claim he was pursuing the origins of his great grandfather's business. Following the departure of Basil from Moore Brothers in 1919, the remaining brothers, Horace and William, invested during 1920 in a new Model T Ford van, HK7284, which was registered as a 'public conveyance'. This indicates an intention to operate a carriers business in competition with Moore and Doughton. However, within three months the van had been sold suggesting family pressures persuaded the brothers this was not the thing to do!

Although Basil was now well into the motor era he never drove any of the vans himself but left this to others. The last driver was Cyril Langstone who lived in Kelvedon. The van continued to operate regularly for some years but by 1944 it disappeared off the scene and so brought to an end the carriers side of the business. Basil died in 1949.

Chapter 5

The Steam Bus

Whilst the steam bus was not the first motor vehicle to become part of the Moore fleet, it is being treated separately because the purchase of this was not only an important landmark in the firm's history, for it was their first double-decker, but it was also an important event for Thomas Clarkson the builder, as will be shown later.

Thomas Clarkson was born in Salford on 24th September 1864. In childhood he possessed a natural interest in things mechanical, and on reaching college opportunities arose for him to become engaged in experimental work. This showed he had a great flair for inventiveness which later led to a number of his patented designs being incorporated in his steam bus chassis. Thomas Clarkson was, without doubt, a very talented designer and this together with his passionate belief in steam as a source of motive power for road vehicles, led him to develop successful steam chassis. In his day, Thomas Clarkson was credited with being 'the most talented designer of steam driven road vehicles that this country has produced'.

His first designs concerned steam cars, but in 1902 he moved to Chelmsford where he set about developing a heavier steam chassis, originally intended for lorries but later for bus use. The premises were situated in Moulsham Street at the junction with Queen Street; the site having been vacated by Colonel Crompton who had now moved to Writtle Road. The foundry was also on this site.

In designing a steam bus Thomas Clarkson had a number of problems to overcome such as the need to produce steam quickly to cope with rapid starting, varying passenger loads and climbing steep hills. Whilst this can be achieved with a large boiler having plenty of reserves of steam, the design is not suitable for a bus where heavy construction needs to be avoided; a bus coming along looking like a traction engine would not have been a pretty sight — at least not at that time! Clarkson overcame this by making use of a lightweight 'flash boiler' which contained a relatively small amount of water and therefore was able to produce

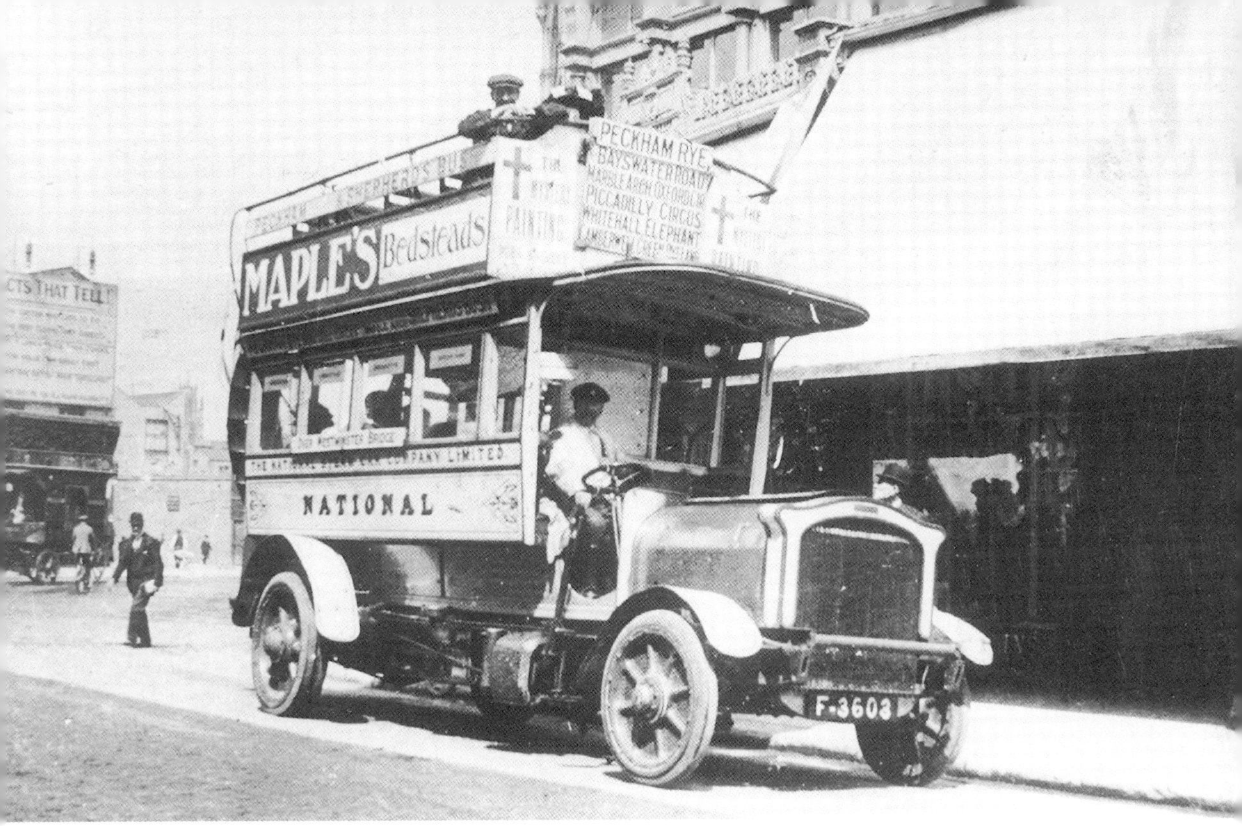

16 Clarkson steam-bus in London.

'instant' steam at the moment it was needed. To do this obviously required the heat output to be finely controlled, as also must be the water feed to the boiler. Paraffin was used as a fuel to heat the boiler. All this was quite a remarkable achievement; for not only was Clarkson responsible for the design but also the manufacture of the necessary component parts.

Buses started to roll off the production line in 1903, and in 1905 Clarkson exhibited his first double-decker. Production and sales of these buses continued in the years ahead and in 1909 Thomas Clarkson decided he was going to operate his steam buses in London under the fleet name 'National' — thus The National Steam Car Company was born, see Pic. 16. During this period (1905) the Great Eastern Railway started to operate buses in the Chelmsford area by using vehicles manufactured entirely in their Stratford works, see Pic. 17. However, by 1913 these buses were reaching the end of their lives and in July of that year The National Steam Car Company took over the operation of these services. Routes were operated to Writtle, Danbury and Great Waltham — Pic. 18 shows a bus operating on one of these routes.

17 GER bus at Chelmsford Railway Station ready to leave for Great Waltham.

18 Clarkson steamer approaching Baddow Road corner, Chelmsford from Danbury.

Meanwhile, the growth of services in London had been very considerable and so had the manufacture of steam buses to meet this demand. In 1914 an operating agreement was reached with London General who was, by far, the largest bus operator in the Capital and able to exercise considerable influence. Initially The National Steam Car Company benefited from this agreement, but it did impose a limit on the size of the fleet in London which had the effect of reducing production output of steam buses in Chelmsford. Therefore, the arrival of an order from Moore Bros. of Kelvedon for a double-decker steam bus must have been like sweet music to the Clarkson Company since it meant that one of their buses would now be operating in North East Essex, with the possibility of orders from other operators in the area, and even over the border from Suffolk.

The bus was delivered in April 1914 and interestingly was painted in the livery of the Steam Car Company, even to the extent of having the fleet name 'National' emblazoned on each side. This suggests the vehicle was built for The National Steam Car Company but at the last moment diverted to fulfil the Moore's order. The main body colour was white with varnish being applied to exposed woodwork and the undercarriage was painted grey. Most probably the body was supplied by Hurst Nelson who, at that time, were well known body builders in Scotland. The public and passengers would often refer to this bus as 'Moore's-National'; which in later years would become a contradiction in terms! Even the timetable of the period refers to 'Moore Bros. National Steam Car'. One can almost imagine the glint in Thomas Clarkson's eyes on knowing this.

Having made the sale, Thomas Clarkson was obviously determined to maximise publicity of this event and arranged a photographic session on Market Hill, Coggeshall. Here we see in Pic. 19 a bus from each company, both apparently operating on stage carriage service but this was portrayal only, since the picture was taken on 8th April 1914 and it is unlikely Moore's had started their service to Colchester by that date. Moreover, there is no record that The National Steam Car Company was operating to Coggeshall at that time. This picture is quite well known and some readers may question whether it really is a publicity shot in view of the disorder of the participants and general informality of the occasion.

However, more recently another picture, Pic. 20, has come to light which shows without doubt the photographic event was well planned with everyone now well organised and obeying the cameraman's order of 'please look at the camera'. Clearly the previous photograph was the result of a trial run. The two people alongside the Moore's bus are William and Basil Moore and the driver is Bob Richardson, the first driver of the Moore's steam bus.

19 Clarkson steamers, National Steam Car and Moore's - Market Hill, Coggeshall, April 1914.

20 As Pic. 19 but now the occasion is formal.

21 Moore's steam bus at their premises 1914. Beside bus:- Basil, Horace, Edward and William Moore. Driver - Bob Richardson.

And so with Clarkson having secured their publicity shot, Moore Bros. were not content with having their new possession photographed beside another operator and decided on a photographic session of their own at their premises in Kelvedon. The result of this session is shown in Pic. 21.

The introduction of the steam bus on the Colchester service was quite an event and the vehicle has been given the credit of being the first passenger-carrying self-propelled bus to run into Colchester from the surrounding district. The silence and smooth running of the bus must have been very impressive to the public at large, compared to the noisy internal combustion engines of other operators in the area. Witnessing a steam bus travelling at night must also have been an awesome sight with its blue glow from the burner jets, but also a great comfort to those waiting at the roadside for a bus knowing it was 'theirs' coming along in the distance. Even so, all this gave little comfort for the bumpy ride ahead at a maximum speed of 12 m.p.h. when passengers could expect to be severely shaken by the solid tyres and the rough road. Nevertheless, we know the bus was often fully loaded with passengers on the Tiptree route and to travel on the steam bus was regarded as quite an experience, as so eloquently expressed at the time by a Miss Blythe. Miss Blythe worked with children at the Sunday School and Band of Hope and composed a poem for them which expressed 'the experience' admirably. The poem can be seen below.

Fig. 8 Miss Blyth's poem.

ABC OF THE TIPTREE BUS.

A for anticipation of Colchester Town
B for the Bus of Tiptree renown
C for the crush which sometimes is bad
D for the dust which makes us so mad
E for the engine which goes at "some" pace
F for the fumes unpleasant to face
G for the gossip which goes on inside
H for the hats you crush on the ride
I for the infants who never keep still
J for the jolts on Heckford Bridge Hill
K stands for Kelvedon place where we start
L for Layer Marney a small rural part
M for "Moore Bros" who take all the tin
N for "No room unless you are thin!"
O for the outside which truly is grand
P for the patience required when you stand
Q for the queue which waits at the store
R for the rush when the Bus starts at four
S for Shrub End with a clock on the church
T for the Talking which gets worse at Birch
U for umbrellas you never can find
V for the victims who get left behind
W for the windows through which you can glare
X for the exit by way of a stair
Y for the yearning to go once again
Z for our zest which never shall wane.

D. C. BLYTH

The steam bus ran the Kelvedon to Colchester service for some years, via Tiptree on Tuesday and Friday and via Coggeshall on Wednesday and Saturday. Timetables from this period can be seen in Fig. 9. It was also available for private hire work and a good example of this is the annual outing of employees from Pinkham's of Witham who were well known

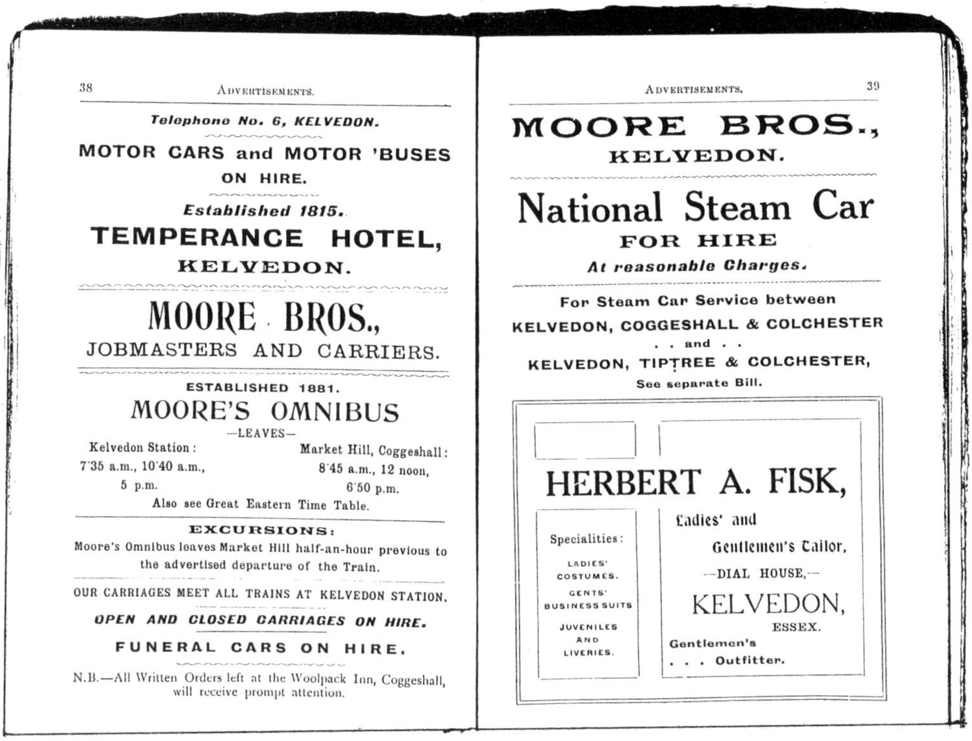

38 ADVERTISEMENTS.

Telephone No. 6, KELVEDON.

MOTOR CARS and MOTOR 'BUSES
ON HIRE.

Established 1815.
TEMPERANCE HOTEL,
KELVEDON.

MOORE BROS.,
JOBMASTERS AND CARRIERS.

ESTABLISHED 1881.
MOORE'S OMNIBUS
—LEAVES—

Kelvedon Station: 7'35 a.m., 10'40 a.m., 5 p.m.
Market Hill, Coggeshall: 8'45 a.m., 12 noon, 6'50 p.m.
Also see Great Eastern Time Table.

EXCURSIONS:
Moore's Omnibus leaves Market Hill half-an-hour previous to the advertised departure of the Train.

OUR CARRIAGES MEET ALL TRAINS AT KELVEDON STATION.

OPEN AND CLOSED CARRIAGES ON HIRE.

FUNERAL CARS ON HIRE.

N.B.—All Written Orders left at the Woolpack Inn, Coggeshall, will receive prompt attention.

ADVERTISEMENTS. 39

MOORE BROS.,
KELVEDON.

National Steam Car
FOR HIRE
At reasonable Charges.

For Steam Car Service between
KELVEDON, COGGESHALL & COLCHESTER
. . and . .
KELVEDON, TIPTREE & COLCHESTER,
See separate Bill.

HERBERT A. FISK,
Specialities: LADIES' COSTUMES. GENTS' BUSINESS SUITS. JUVENILES AND LIVERIES.
Ladies' and Gentlemen's Tailor,
—DIAL HOUSE,—
KELVEDON,
ESSEX.
Gentlemen's . . . Outfitter.

Fig. 8a Advert from local Almanack, 1915.

glove makers. A picture of the start of the outing is shown in Pic. 22. Both the date and destination are unknown, but it was probably taken around 1915 going by the condition of the vehicle. Unfortunately contact with the Pinkham family produced little additional information, but we do know the gentleman in the bowler hat is Mr W. Pinkham with his little dog 'Curly'. On the left of Mr Pinkham is his wife Rebecca and to his right his son Leslie. An article on this photograph was featured in the 'Braintree and Witham Times' some years ago.

A driver of a steam bus had to have a sound knowledge of steam operation and then he would be classed as a 'Steam Engineer'. One such person was Alfred Pentney who joined the firm in April 1919 taking over from James Waldon. Fortunately for us he kept a diary for the first few months of his employment with Moore Bros. and this gives a good insight into driving and working with a steam bus. David Thornton in his book 'Just A Ticket' quoted from a number of entries in this diary and there are just two incidents from the diary which I would like to relate here.

The first one creates a strange irony in that Alfred Pentney arranged for the ex-A.W. Berry's Thornycroft, which Moore's purchased in May 1919, to be accommodated in the garage and the following day for the operator's name to be changed. Little did Alfred Pentney know that the life of the body on the Thornycroft was to be short-lived and would be replaced with the body from his steam bus when it was withdrawn from service. However, even after removal from the Thornycroft some years later, the body was to serve another life when it became a shed on Archie Pentney's smallholding opposite Bob's Cafe in London Road, Kelvedon. The dropped kerb leading to the field can still be seen. Here it was to remain for many years before being broken up around 1950. Pic. 23 shows the body in a very dilapidated state about this time.

The second incident concerned Alfred Pentney getting into trouble with Colchester Police on two occasions within a day or two of each other. The first time was due to the bus breaking an electric lamp in Lexden Road, Colchester, when he was later questioned by a Police Inspector in Kelvedon about the incident. Moore Bros. received a letter about the

MOORE BROS.'
KELVEDON,
NATIONAL STEAM CAR
WILL RUN
Every Tuesday and Friday
From KELVEDON to TIPTREE and COLCHESTER.

TIMES AND FARES AS UNDER:

LEAVE	a.m.	LEAVE	p.m.
Kelvedon	11·10	Colchester Town Hall	4·30
Inworth Street	11·20	Birch Rectory Corner	5·0
Tiptree Co-operative	11·30	Smythes Green	5·5
Haynes Green Corner	11·35	Haynes Green Corner	5·10
Smythes Green	11·40	Tiptree Co-operative	5·20
Birch Rectory Corner	11·50	Inworth Street	5·25
ARRIVE		ARRIVE	
Colchester Town Hall	12·15	Kelvedon	5·35

FARES:

	Adults	Children
Kelvedon to Tiptree	4d.	2d.
Inworth Street to Colchester	11d.	6d.
Tiptree to Colchester	9d.	6d.
Haynes Green Corner to Colchester	8d.	5d.
Smythes Green to Colchester	7d.	4d.
Birch Rectory Corner to Colchester	6d.	4d.
Kelvedon to Colchester	1/-	8d

SPECIAL TRIPS ARRANGED.
This Car stops to pick up and set down Passengers anywhere.

Return Fare, Kelvedon to Colchester either by Tiptree or Coggeshall, 1/9, Children, 1/2.
Children under 5 years, free,

J. A. Dodds, Coggeshall Press.

MOORE BROS.'
KELVEDON,
NATIONAL STEAM CAR
WILL RUN ON
Wednesdays and Saturdays
From Kelvedon to Coggeshall and Colchester.

TIMES AND FARES AS UNDER:

LEAVE	a.m.	LEAVE	p.m.
Kelvedon	11·10	Colchester Town Hall	4·30
Pound Farm	11·15	Marks Tey School	4·55
Hamlet	11·20	Elm Farm, Colchester Road	5·5
Coggeshall	11·30	Surrex	5·10
Surrex	11·40	Coggeshall	5·20
Elm Farm, Colchester Road	11·45	Hamlet	5·25
Marks Tey School	11·55	Pound Farm	5·30
ARRIVE		ARRIVE	
Colchester Town Hall	12·20	Kelvedon	5·35

FARES:

	Adults.	Children
Kelvedon to Coggeshall	3d.	2d.
Coggeshall to Colchester	9d.	5d.
Surrex to Colchester	8d.	5d.
Elm Farm to Colchester	7d.	4d.
Marks Tey School to Colchester	6d.	4d.
Coggeshall to Marks Tey	5d.	3d.
Kelvedon to Colchester	1/-	8d.
Pound Farm to Colchester	11d.	7d.
Hamlet to Colchester	10d.	6d.

SPECIAL TRIPS ARRANGED.
This Car stops to pick up and set down Passengers anywhere.

Return Fare, Kelvedon to Colchester either by Tiptree or Coggeshall, 1/9, Children, 1/2.
Children under 5 years, free.

Fig. 9 1914 timetables.

22 Steam bus and Thornycroft charabanc outside Pinkham's factory Witham c1915.

damage, when 7s.6d. was required to pay for a replacement. Alfred Pentney had to take the money to Colchester Town Hall on the Friday, but on Monday he was again in trouble with the Colchester Police for taking water in the High Street for the bus boiler, as he didn't pay water rates in the Borough! What resulted from this is not recorded.

From Alfred Pentney's diary whilst it is evident that some of the stage carriage and private hire work in 1919 was being taken on by the ex-Berry Thornycroft double-decker, the steam bus was still carrying a heavy schedule of stage carriage work, as well as private hire duties, to such places as Southend-on-Sea, Dovercourt and Bury St Edmunds. It is believed the steam bus was withdrawn from service about 1921 and, apart from the body, was probably broken up.

There was talk of Moore Bros. buying another Clarkson steamer during 1919 but nothing became of this. This was a wise move since steam propulsion for buses was nearing its end with the increased reliability of internal combustion engines and as we have already seen, Moore's first petrol engined double-decker was a second-hand Thornycroft from A. W. Berry. The firm must have been well satisfied with this make since they went on to purchase more of these ex-war department subsidy J type chassis, which were then fitted with second-hand double-decker bodies.

Clarkson steam bus manufacture ceased towards the end of 1919, when AEC petrol engined buses were introduced into the fleet of The National Steam Car Company. In February 1920 the operating name was changed from The National Steam Car Company to The National Omnibus and Transport Co. Ltd., from which descended The Eastern National Omnibus Co. Ltd.

23 Steam bus body in field, London Road, Kelvedon.

Chapter 6

The Roaring Twenties

Roaring they most certainly were. With surplus army vehicles available, there were plenty of would-be busmen to take the plunge into the business of passenger carrying. Because some saw this as seasonal opportunities the chassis were adapted to accept bus or charabanc bodies in summer, and lorry bodies in winter. Some bus proprietors who had been running regular services well before this decade suddenly found newcomers appearing and operating just a few minutes ahead of their published schedules. Moreover, the established proprietors were generally licensed with the local authorities whereas the new arrivals often were not. The 'pirates' as they became known, employed various and devious tactics to gain passengers and so fill their buses. It was not uncommon for a pirate to pass by a solitary person waiting for a bus, in

HACKNEY CARRIAGE LICENCE.

No. ________

BOROUGH OF COLCHESTER.

We, the Mayor, Aldermen and Burgesses of the Borough of Colchester, being the Urban District Council for the District of the said Borough, Do hereby License ________ of ________ to ply for hire within the said Borough, and to carry ________ persons, in conformity with the provisions of the Statutes in that behalf, and the Byelaws for the time being in force in the said Borough, relating to Hackney Carriages, the Hackney Carriage numbered ________, and having ________ wheels, called a ________

The Common Seal of the Mayor, Aldermen and Burgesses of the Borough of Colchester, as and acting as such Urban District Council as aforesaid, was affixed hereto in pursuance of a resolution passed by a meeting of the Council of the said Borough, duly convened and held on the ________ day of ________ One thousand nine hundred and ________ and the same was so affixed, in the presence of

________ *Mayor.*

________ *Town Clerk.*

NOTE.—This Licence will be in force only until 1st day of May, 19

The two numbered enamelled plates, each of which hears a number corresponding to that specified on this Licence, which are delivered to the Licencee, are the **sole property** of the said Mayor, Aldermen and Burgesses, and are merely **LENT** to the Licencee for the period during which this Licence is in force, and for no longer, and must he delivered up to the Inspector of Hackney Carriages at the expiration, or sooner determination, of this Licence.

600-6/15

Fig. 10 Borough of Colchester Hackney Carriage Licence.

order to pick up multiple fares at a stop further along the route. Apart from all this, the buses were often badly maintained giving rise to breakdowns, even to the extent of the operator not running a service at all on a particular day. Nevertheless the cutting of fares made travelling on these buses attractive to some people.

Clearly a large number of the established operators who were trying to run regular and reliable services found these illicit practices very upsetting. Although local authorities were given the powers to licence hackney carriages plying for hire within their boundaries, under the Public Health Act 1875 and the Towns Police Clauses Acts 1847 and 1889, these powers were not always applied at the start of motor bus operation. Some who did adopt these powers applied them unfairly whereas others lacked enforcement. In some instances, such as Grays in Essex, the authority did not take up licensing of buses at all, with chaotic results due to the large influx of vehicles. However, the main towns with which we are concerned in our story, namely Colchester and Chelmsford, established licensing for horse drawn carriages and later for motor buses. Whilst Moore's applied to Chelmsford Town Council for driver and hackney carriage licences in 1905 their applications were not as numerous as those applied for from the Borough of Colchester, presumably because the firm had foreseen there were greater opportunities for expansion in Colchester than in Chelmsford. An example of a Borough of Colchester application form for a hackney carriage licence can be seen in Fig. 10. From this it will be seen that badge numbers were allocated. These numbers had to be affixed to the rear of the vehicle and generally took the

MOORE BROS., Motor Contractors and Saloon Bus Proprietors, **KELVEDON.**

Bus Park, High Street, Hospital and Lexden.

Lexden (Borough Boundary).												
1	**Lexden** (Straight Road).											
1	1	**Lexden,** King's Head.										
1½	1	1	**Lexden Church.**									
2	1½	1½	1	**Fitzwalter Road.**								
2	1½	1½	1	1	**Park Road.**							
2½	2	2	1½	1	1	**Beverly Road.**						
2½	2	2	1½	1	1	1	**Hospital.**					
3	2½	2½	2	1½	1½	1	1	**Head Street.**				
4	3½	3½	3	2½	2½	2	1½	1	**High Street.**			
4	4	4	3½	3	3	2½	2	1	1	**All Saints Church.**		
4	4	4	4	3½	3½	3	2½	1½	1½	1	**St. Botolph's.**	
4	4	4	4	4	4	3½	3	2	2	1½	1	**Bus Park.**

Children between 3 and 12, Half Fare.
No Fare less than ½d. and fraction of ½d. charged as ½d.
School Children going to and from School charged 1d. from Head Street, and anywhere in the Borough.

S. G. PARISH, COGGESHALL PRESS.

Fig. 11 Fares, within Colchester Borough boundary.

24 c1924 construction St. John's Bus Station, Colchester.

form of an enamelled plate, Colour Pic. 1. The first Moore's bus to receive a licence was the steam bus. Licence No. 128 was issued on 3rd June 1914 and as we have seen in Chapter 5 it was the first motor bus to enter Colchester from an outside district. Drivers and conductors also had to be licensed, the numbers being carried on enamelled badges hung from a

25 Departure board, St. John's Bus Station, Colchester.

26 c1928. Line up of buses St. John's Bus Station two Moore's buses can just be seen a Reo and a Chevrolet.

leather strap attached to a button on the jacket or coat. The badge number shown in the same colour picture belonged to Toby Axten who was a conductor at that time with A.W.Berry.

It should be noted that local authorities often dictated the fares that were to be charged by operators within the authorities boundaries. Generally these were issued to protect their own transport undertakings, such as in the case of Colchester Corporation Transport. In Fig. 11 can be seen a fare table on the Lexden route into Colchester in conformity with Colchester Corporation's fare structure, but published by Moore's. At one time special tickets were introduced by Colchester Corporation for use by other operators within the boundary.

All buses into Colchester terminated and departed from a point outside the Hippodrome in the High Street. Here the road was at its widest allowing both buses and taxis to stand, although buses were only permitted to stand for 15 minutes before departure time. Layover periods made use of Stanwell Street for parking until the 'Colchester Parking Station', as it was then called, was opened in 1926 in St. John's Street. A picture showing the station under construction is shown in Pic. 24, and it is interesting that a booking office and a bench seat for passengers, belonging to Wormell's Motor Coaches is still in place, but shortly would be demolished. A temporary departure board can be seen alongside St. John's Street showing that some operators, including Moore's, were already using the parking station before completion. The huge departure timetable board which was situated at the east end of the station is shown

27 Ford T charabanc en route Clacton 1925. Driver is George Buck.

in Pic. 25. A close look at this picture suggests the board was erected during early construction since a number of buses are facing into the station having entered from St. John's Street. It would seem that the entrance via Abbeygate Street had yet to be opened up, allowing buses to draw up facing St. John's Street in the all too familiar fashion and in the stands dictated by the departures board. Pictures taken later in 1928 show the station being used in the way it was planned and this continued until the new bus park opened in The Minories in 1961, see Pic. 26.

We now need to turn our attention to the way in which the Moore's bus fleet was expanding, as well as looking at the growth in passenger traffic. The acquisition of the Thornycroft from Berry's was a success which led to more purchases of Thornycroft open-top double-deckers. Also at this time a number of Model T Ford chassis were bought new which had 14 seater bus and charabanc bodies fitted. These smaller capacity vehicles were not only required for the expansion of stage services, but also to support the ever increasing trips to the seaside resorts during the summer months. A Model T Ford charabanc on a seaside outing in 1925 is shown in Pic. 27.

By the early 1920's a service to Colchester via Easthorpe and Copford Green had been started but on Fridays only, and a Model T Ford operating this service can be seen in Pic. 28. About this time a service from Braintree to Colchester was opened up on Wednesdays only with one journey in each direction. We also find a main road service between Kelvedon and Colchester operating on Fridays only but with a late night trip in each direction on Saturdays. However, by the end of the decade most of these services had developed into daily services with greatly increased frequencies. And finally by 1930 a daily through service from Colchester to Chelmsford had begun by extension of the Kelvedon to Colchester main road route. It is possible this latter service was hastily put into operation due to rumours circulating that those operators who did not have services operating by 1930 would be unlikely to get licenses allocated with the coming of the Road Traffic Act 1930.

28 August 1924. Ford T at Feering on Kelvedon -Colchester via Easthorpe service. Driver is Jimmy Pitchford with Doreen who became a conductress WW II.

During the early 1920's a form of issuing fare tickets was introduced on Moore's buses. These were of a primitive nature and partly made out in pencil by the conductor at the time of issue. To my knowledge none of these has survived. Later on a printed version became available and three denominations of these tickets, together with a Williamson version from approximately 1927 have survived. In 1928 Moore's converted to the Bell Punch system which lasted until it was replaced by the insert version of the Setright Registers in about 1950. A range of tickets covering the whole of this era is shown in Colour Pic. 2. Ticket racks and machines from both these systems, also a moneybag, are displayed in Colour Pic. 3.

28a Fares please Ted!

TED TAKES THE BUS TO COLCHESTER

Ted, the Collie dog who belonged to William Moore, was loved by all the children in Kelvedon and was famous for frequently taking the Moore's Thornycroft to Colchester. The Essex County Telegraph of 1927 featured him in an article because he was renowned for never getting on the wrong bus and always took the one going by the direct route.

28b Come on - lets go!

29 Reo Sprinter. Angel, Kelvedon c1930.

30 GMC on Mothers Union outing Windsor 1929.

As we draw to the end of the decade we find the vehicle fleet changing quite significantly. Although the Thornycroft chassis continued to find favour for larger capacity carrying, the Model T Ford lost ground to other manufacturers. No more purchases of Fords were made after 1924 — the last one being a van, the subject of which was covered in Chapter 4. Although the Model T Fords remained in the fleet up to 1930 gradually they were replaced with makes such as Chevrolet, GMC and REO. All the Chevrolets were small capacity saloons having 14 seats only, but the

31 Reo, Angel Hill, Bury St. Edmunds c1930.

GMC's and REO's carried a range of bodies from 20 seaters to 32 seaters. Examples of these vehicles in service are shown in Pics. 29, 30 and 31. As we move into the 1930's we will see that yet other makes will join the fleet giving even greater variety of chassis and body manufacturers on which to travel. Not all purchases would be new however, although it would be company policy to buy new stock whenever possible.

31a Thornycroft accident at Frating. 1923.

The following appeared in the East Anglian Daily Times of July 27 1923

'BUS IN A DITCH.

ALARMING EXPERIENCE ON CLACTON ROAD.

Whilst a motor bus belonging to Mr. Moore, of Kelvedon, and conveying a large party of holiday-makers on Thursday morning from Tiptree to Clacton, was near Frating, the hub of the offside front wheel of the 'bus was struck by an overtaking Ford car, and the steering wheel was knocked out of the 'bus driver's hands.

The 'bus plunged into a ditch, with its rear wheels in the air, and several outside passengers were thrown off into a field, one of them turning a complete somersault.

Several sustained cuts and bruises, the most seriously hurt being an elderly lady named Hart, of Tiptree, who was internally injured.

Dr. Athill, of Great Bentley, was summoned, and was promptly in attendance.

The passengers were chiefly from Tiptree and Birch, and although a large number of them were women and children, there was no panic or screaming when the accident occured. The passengers who were unhurt assisted those suffering from cuts and contusions, and Mr. H. K. Moore, who was driving and who had an almost miraculous escape when the 'bus rolled on its side as the forepart went into the ditch, which was a deep one, broke one of the large windows in order to release the inside passengers. The front axle and wheels were cut clean from the body of the vehicle, which otherwise sustained very little damage, although one of the back seats was smashed off.

Mr. E. H. Andrews, Superintendent of the St. John's Ambulance Corps, was telephoned for and arrived at Frating in quick time with the Colchester Ambulance waggon, but fortunately, there were no stretcher cases to deal with, and his services were not required.

Chapter 7

The Service Network and Bus Regulation

The Road Traffic Act 1930 was a very significant piece of transport legislation. The part which related to Public Service Vehicles (PSV) came into force on 1st April 1931 and had an immediate impact on the passenger carrying industry. Any vehicle having eight or more seats and engaged on passenger carrying would be subject to the Act and would require to be licensed whether on stage, express (including excursions and tours), or contract carriage. A typical example of such a licence appears in Pic. 62 in Chapter 12 and this would be required in addition to the normal road fund licence. Furthermore, to operate a service a 'Road Service Licence' would be required and the application would need to include route details, timetables, fare structures, as well as the maximum number of PSV vehicles to be employed. All applications had to be submitted to the area office of the Traffic Commissioners in which the services were to be operated and if the service meant entering another traffic area then a 'backing licence' would be required from that area office. The application forms asked for considerable detail, also a section had to be completed to a format suitable for direct inclusion in the N&P.

Lists of applications, dates of hearings in the traffic courts and the decisions resulting from these, were given in regular issues of 'Notices and Proceedings' (N&P's) issued by the Traffic Commissioners. A list of Moore's initial applications for Road Service Licences as they appeared in the N&P's, is given in Appendix III together with the Traffic Commissioners decisions on grants or refusals for these applications. At the beginning licences were granted for periods of 6 to 12 months only, involving frequent renewals and much administrative effort. Later on licences were generally valid for 3 years.

No departure from the terms of the issued licence was permitted, and the nervousness with which an operator initially viewed these restrictions can be gauged by the number of 'Traffic Notices' issued by Moore Bros. and displayed on their vehicles and elsewhere during the early period of regulation. These notices were, of course, intended to inform passengers of changes in services, but because of their formal nature also served to show to the Traffic Commissioners the firm's eagerness to conform to recent legislation. Any operator wishing to vary the conditions of a licence did so by making an application using a 'Continuation of Service with Modification' form. Procedures would then follow the normal course of appearing in N&P's under: application, notice of hearing (with period then allowed for objections from other operators), granting or refusal. Examples of applications and grants as they appeared in the N&P's are given in Appendix IV.

We have seen how local authorities, Colchester included, had introduced even before 1930 a form of regulation, but this was limited in effect. Those regulations administered by the Traffic Commissioners under the 1930 Act were extensive and any operator contravening his licence was liable to have it revoked. Road Service Licences were not freely issued and generally it was only those who had been running services before 1930, had held licences from local authorities where appropriate, and possessed a good service record, who would receive a licence. A further point was that the initial applications had to be made by January 1931.

The Traffic Commissioners were also responsible for the issue of licences to drivers and conductors. Licences had a life of three years and re-application was necessary after this time. These licences were issued with stringent conditions attached and any contravention of these could result in a fine and endorsement of the licence. Misconduct could mean the licence being suspended or even revoked. Minimum age limits applied to these licences, 21 years for drivers and 18 years for conductors. Examples of these licences, together with the familiar red and green badges which had to be worn by all drivers and conductors, are shown in Colour Pic. 1.

All buses were subject to regular inspection by the Ministry of Transport PSV Examiner and a 'Certificate of Fitness' issued, if approved. Presentation of a vehicle was very important, in that the underside had to be free from mud and grease so the examiner could do his job thoroughly. If it were not, the examiner could instruct that the bus be withdrawn from service. This could also apply if serious defects to body and chassis were discovered, probably resulting in a costly exercise for the operator. Certificates were issued for a period of between 3 and 5 years, dependent on age and condition of the vehicle. On average a Certificate had a life of 3 years. However, if a spot check was carried out and serious defects were found, the Certificate could be revoked. Otherwise a time limit would be given for repairs to be carried out. An example of a Certificate is shown in Pic. 62 in Chapter 12.

Local Authorities already had the power to determine the picking up and setting down points for passengers. This brought to an end Moore's long standing claim in their publications, that 'this CAR (bus) stops to pick up and set down PASSENGERS anywhere'. Such powers were applied by the Borough of Chelmsford, amongst others, and their regulations regarding routes, stopping places and termini within the Borough were given in a letter dated 22nd August 1930, of which the following is an extract:-

Omnibus Terminus

(a) That omnibuses be prohibited from stopping and starting, for the purpose of setting down or picking up passengers in Duke Street between the National Garage and Market Road, and that on entering the town (with the exception of the Eastern National Omnibus Company's vehicles, which will run to and from their own garage), they turn down Market Road from Duke Street, set down and pick up their passengers by the wall opposite the Eastern Garages and leave again by Market Road, except Fridays, when they will leave by King Edwards Avenue.

(b) That no omnibuses shall draw up or remain on the "Terminus" more than five minutes before the scheduled time of departure.

(c) That omnibuses which have to wait any length of time before leaving on their next service shall park in the Car Park at rear of Market, or in the car-standing places in Coval Lane or Victoria Road.

Omnibus Stopping Places.

The following to be recognised as omnibus stopping places:-

(a) Omnibuses entering the town :-

From Springfield	Luckin Smith's Ironmongery Shop
	Tindal Statue
	Terminus, Market Road

(b) Leaving the town :-

To Springfield	Terminus, Market Road
	Cattle's Shop near Golden Lion
	Opposite Luckin Smith's Ironmongery Shop
	Opposite Three Cups Hotel
	Post Office, Springfield Hill

With the growth of motor traffic, more places were making use of these powers and stopping places in Kelvedon were announced by the Traffic Commissioners in N&P No. 368 dated 18th July 1938, detailed as follows:-

1. Vehicles operating from Colchester to Chelmsford.
   (1) Swan Street Corner, adjacent to the Railway Tavern.
   (2) High Street, Kelvedon. (Tail stop at telegraph pole outside Rowe's shop.)
   (3) High Street, Kelvedon. (Outside the Masonic Hall.)
   (4) Maldon Road Corner. (At the south-west corner of the existing island.)
2. Vehicles operating from Chelmsford to Colchester.
   (1) High Street, Kelvedon. (Head stop outside "The Hollies".)
   (2) High Street, Kelvedon. (Tail stop at telegraph pole opposite Infants' School.)
   (3) High Street, Kelvedon. (Tail stop at telegraph pole outside Courtney House.)
   (4) Station Road Corner. (At the electric light pole about 10 yards from the Corner House Café.)

Passengers may be taken up and set down outside the limits of the prescribed stopping places at any point en route where traffic conditions permit.

Where stopping places had been imposed, regulation size 'Bus Stop' signs started to appear from 1933 onwards.

A number of the travelling public viewed these events as draconian and regarded them as not being in their best interests. They feared all this would lead to their local friendly bus operator disappearing off the scene. To some extent this fear was whipped up by the small operators before the Act came into force. Also, some operators were frightened of loosing their livelihoods, particularly those who ran their businesses on a shoestring with little regard for legislation and formality. In some quarters opposition to the Act was intense and the travelling public were encouraged to protest by writing to their Members of Parliament at the House of Commons. As an inducement, printed cards were circulated and one such example is shown in Fig. 12.

When the 1930 Act came into force a number of operators had their worst fears realised when some applications were turned down, while others

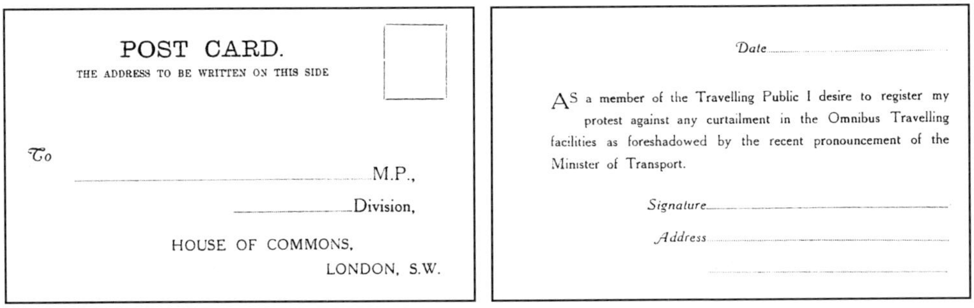
POST CARD.

THE ADDRESS TO BE WRITTEN ON THIS SIDE

To

M.P.,

Division,

HOUSE OF COMMONS,

LONDON, S.W.

Date

AS a member of the Travelling Public I desire to register my protest against any curtailment in the Omnibus Travelling facilities as foreshadowed by the recent pronouncement of the Minister of Transport.

Signature

Address

Fig. 12 Registration of protest card.

didn't even bother to apply and withdrew their bus services. For those who had lost out but had the determination to succeed, there was always the appeal procedure and legal representation was allowed if the means were available. At the end of the day independent operators, both large and small, did survive and prosper, as the Moore Bros. story gives testimony.

It should be noted that the foregoing does not cover all the implications of the 1930 Act, only those items which have a significant bearing on our story have been included. It is worth noting that the main elements of regulation contained in the 1930 Act continued in force throughout the remaining life of Moore's operations. Indeed regulation continued until Margaret Thatcher's government brought in deregulation through the Transport Act 1985, which came into force in October 1986. A much diluted form of regulation now applies.

Once operators had secured their Road Service Licences it was inevitable they would try to resist newcomers appearing on their routes. Any intending operator would need to apply for a licence and give reasons why such a service was needed. As we have seen existing operators would be made aware of the application through N&P's issued by the area Traffic Office, which for the Eastern Area was, and still is, at Cambridge. Fourteen days were allowed for objections and if any were received the application would be heard in a traffic court with the applicant being cross-examined by existing operators. This meant time, delay, and expense, and then possibly some, or all, of the licences being refused. For this reason, many bus firms sought expansion through buying up operators businesses and/or acquiring their licences. This was particularly the policy of combines such as Eastern National who were part of the Thomas Tilling group.

It was in this way that Moore's secured their Maldon route to Chelmsford (via Hatfield Peverel), and also the Tolleshunt Knights route to Colchester, through the transfer of licences. These licences were originally granted to Mr A.V.Brown proprietor of Bluebird Motor Coaches of Tiptree, who in 1932 decided to cease running buses. An example of a Bluebird timetable, also a picture of their garage is shown in Pic. 32. Reference to Appendix III will show the application and granting of these licences to Moore Bros. In granting these licences to Mr A. V. Brown in 1931 they had been 'Subject to Co-ordination' and this was not an uncommon condition to be imposed by the Traffic Commissioners. Many operators had entered into working agreements with their competitors prior to 1931 and therefore intervention by the Traffic Commissioners in such cases was not necessary. 'Subject to Co-ordination' had been imposed upon Moore Bros. in their initial applications in one instance

"BLUEBIRD"

MOTOR COACHES

TIME TABLE

V. BROWN, Proprietor, TIPTREE,

ESSEX.

Telephone—Tiptree No. 40.

Maldon, Hatfield and Chelmsford.

Monday to Saturday.

								S.O.	N.S.D.	S.O.D.	D.	D.	S.O.
Depart Maldon.	a.m.	a.m.	a.m.	a.m.	p.m.	p.m.	p.m.	p.m.	p.m.	p.m.	p.m.	p.m.	p.m.
Maldon Ship and Anchor ..	9 0	9 45	10 50	11 35	1 40	2 15	3 20	3 45	5 0	5 30	6 30	7 30	9 20
Heybridge Square	9 5	9 50	10 55	11 40	1 45	2 20	3 25	3 50	5 5	5 35	6 35	7 35	9 25
Langford P.O.	9 10	9 55	11 0	11 45	1 50	2 25	3 30	3 55	5 10	5 40	6 40	7 40	9 30
Hatfield Green	9 20	10 5	11 10	11 55	2 5	2 35	3 40	4 5	5 20	5 50	6 50	7 50	9 40
Hatfield Duke	9 22	10 7	11 15	11 57	2 7	2 37	3 42	4 7	5 22	5 52	6 52	7 52	9 42
Boreham Cock	9 27	10 12	11 18	12 2	2 10	2 41	3 47	4 12	5 27	5 57	6 57	7 57	9 45
Boreham House	9 32	10 17	11 23	12 7	2 13	2 50	3 52	4 17	5 32	6 2	7 2	8 2	9 52
Springfield Plough	9 37	10 22	11 28	12 12	2 17	2 52	3 57	4 22	5 37	6 7	7 7	8 7	9 57
Springfield Road Corner ..	9 40	10 25	11 31	12 15	2 20	2 53	4 0	4 25	5 40	6 10	7 10	8 10	10 0
Chelmsford Station	9 44	10 29	11 35	12 19	2 24	2 55	4 4	4 29	5 44	6 14	7 14	8 14	10 4

Sunday.

								D.	D.	D.	D.	D.
Depart Maldon.	a.m.	a.m.	a.m.	a.m.	p.m.	p.m.	p.m.	p.m.	p.m.	p.m.	p.m.	p.m.
Maldon Ship and Anchor ..	8 40	9 15	10 35	11 0	12 30	1 0	3 0	4 50	5 55	6 50	7 45	9 15
Heybridge Square	8 45	9 20	10 40	11 5	12 35	1 5	3 5	4 55	6 0	6 55	8 0	9 20
Langford P.O.	8 50	9 25	10 45	11 10	12 40	1 10	3 10	5 0	6 5	7 0	8 5	9 25
Hatfield Green	9 0	9 35	10 55	11 20	12 50	1 20	3 20	5 10	6 15	7 10	8 15	9 35
Hatfield Duke	9 2	9 37	10 57	11 22	12 52	1 22	3 22	5 12	6 17	7 12	8 17	9 37
Boreham Cock	9 7	9 42	11 2	11 27	12 57	1 27	3 27	5 17	6 22	7 17	8 22	9 42
Boreham House	9 12	9 47	11 7	11 32	1 2	1 32	3 32	5 22	6 27	7 22	8 27	9 47
Springfield Plough	9 17	9 52	11 12	11 37	1 7	1 37	3 37	5 27	6 32	7 27	8 32	9 52
Springfield Road Corner ..	9 20	9 55	11 15	11 40	1 10	1 40	3 40	5 30	6 35	7 30	8 35	9 55
Chelmsford Station	9 24	9 59	11 19	11 44	1 14	1 44	3 44	5 34	6 39	7 34	8 39	9 59

Monday to Saturday.

	D.	D.	D.	D.	D.	D.		S.O.			S.O.	N.S.	S.O.	S.O.
Depart Chelmsford.	a.m.	a.m.	p.m.	p.m.	p.m.	p.m.	p.m.	p.m.	p.m.	p.m.	p.m.	p.m.	p.m.	p.m.
Chelmsford Station	10 0	10 45	12 0	1 30	2 30	3 0	4 15	4 45	6 30	7 30	8 30	8 45	9 5	10 30
Springfield Road Corner ..	10 3	10 48	12 3	1 33	2 33	3 3	4 18	4 48	6 33	7 33	8 33	8 48	9 8	10 33
Springfield Plough	10 6	10 51	12 6	1 36	2 36	3 6	4 21	4 51	6 36	7 36	8 36	8 51	9 11	10 36
Boreham House	10 11	10 56	12 11	1 41	2 41	3 11	4 26	4 56	6 41	7 41	8 41	8 56	9 16	10 41
Boreham Cock	10 16	11 1	12 16	1 46	2 46	3 16	4 31	5 1	6 46	7 46	8 46	9 1	9 21	10 46
Hatfield Duke	10 21	11 6	12 21	1 51	2 51	3 21	4 36	5 6	6 51	7 51	8 51	9 6	9 26	10 51
Hatfield Green	10 23	11 8	12 23	1 53	2 53	3 23	4 38	5 8	6 53	7 53	8 53	9 8	9 28	10 53
Langford P.O.	10 33	11 18	12 33	2 3	3 3	3 33	4 48	5 18	7 3	8 3	9 3	9 18	9 38	11 3
Heybridge Square	10 38	11 23	12 38	2 8	3 8	3 38	4 53	5 23	7 8	8 8	9 8	9 23	9 43	11 8
Maldon Ship and Anchor ..	10 43	11 28	12 43	2 13	3 13	3 43	4 58	5 28	7 13	8 13	9 13	9 28	9 48	11 13

Sunday.

	D.	D.	D.	D.	D.	D.	D.					
Depart Chelmsford.	a.m.	a.m.	a.m.	p.m.	p.m.	p.m.	p.m.	p.m.	p.m.	p.m.	p.m.	p.m.
Chelmsford Station	9 45	10 15	11 30	12 0	1 45	2 15	4 0	5 45	6 45	8 30	9 15	10 0
Springfield Road Corner ..	9 48	10 18	11 33	12 3	1 48	2 18	4 3	5 48	6 47	8 33	9 18	10 3
Springfield Plough	9 51	10 21	11 36	12 6	1 51	2 21	4 6	5 51	6 50	8 36	9 21	10 6
Boreham House	9 56	10 26	11 41	12 11	1 56	2 26	4 11	5 56	6 55	8 41	9 26	10 11
Boreham Cock	10 1	10 31	11 46	12 17	2 1	2 31	4 17	6 1	6 56	8 46	9 31	10 17
Hatfield Duke	10 6	10 37	11 51	12 22	2 6	2 36	4 22	6 6	7 3	8 51	9 36	10 22
Hatfield Green	10 8	10 39	11 53	12 26	2 8	2 38	4 24	6 8	7 5	8 53	9 38	10 24
Langford P.O.	10 18	10 49	12 3	12 35	2 18	2 48	4 34	6 18	7 15	9 3	9 48	10 34
Heybridge Square	10 23	10 54	12 8	12 40	2 23	2 53	4 39	6 23	7 20	9 8	9 53	10 39
Maldon Ship and Anchor ..	10 28	10 59	12 13	12 45	2 28	2 58	4 44	6 28	7 25	9 13	9 58	10 44

S.O.—Saturdays only. N.S.—Not Saturdays.

D.—During June, July, August and September, these times will connect up at Heybridge Square with our Bus running to and from Mill Beach.

32 Vic Brown with Essex Car, Reo and Chevrolet. Charlie Sorrell on right later joined Moore's as a driver.

only — which was on the Braintree to Colchester route — clearly showing agreements existed on other services where competition was present.

Attempts were made by Moore Bros. to obtain a service from Hatfield Peverel to Braintree via the Notleys, which was operated by H. Springett and Son of Hatfield Peverel. An example of Springett's timetable and bus ticket is shown in Fig. 13. His premises and garage were at the west end of Hatfield Peverel, opposite The Crown (now the William Boosey) at the top of the hill on the old A12. Mr Springett wanted to sell the business for

Phone: Hatfield Peverel 35.

SPRINGETT'S BUS SERVICE.

DAILY SERVICE (including Sundays)

BETWEEN

Hatfield Peverel, Witham and Braintree

Via FAULKBOURNE and the NOTLEYS.

				WSO	WO			WO	NW		SO		SO	WSO
	a.m.	a.m.	a.m.	p.m.	p.m.	p.m.	p.m.	p.m	p.m.	p.m.	p.m.	p.m.	p.m.	p.m.
Hatfield Peverel Dept.	7.50	9.30		12.15	—	1.50	—	—	3.20	5. 0	6. 0	7. 0	—	9.20
WITHAM	8. 0	9.45		12.25	1.35	2. 0	—	3.30	3.30	5.10	6.10	7.20	—	9.30
CHIPPING HILL	8. 4	9.49		12.28	1.38	2. 3	—	3.33	3.33	5.13	6.14	7.23	—	9.33
VICTORIA, Power's Hall End	8. 6	9.51		12.31	1.41	2. 5	—	3.35	3.35	5.15	6.16	7.26	—	9.36
FAULKBOURNE (P. Office)	8.12	9.57		12.37	1.47	2.12	—	3.42	3.42	5.22	6.22	7.32	—	9.42
WHITE NOTLEY (X Keys)	8.16	10. 1	10.50	12.41	1.51	2.16	3. 5	3.46	3.46	5.26	6.26	7.36	8.16	9.46
SANATORIUM	8.22	10. 7	10.55	12.47	1.56	2.22	3.10	3.52	3.52	5.32	6.32	7.42	8.21	9.58
RIFLE HILL	8.29	10.14	11. 3	12.54	2. 0	2.29	3.17	3.59	3.59	5.39	6.39	7.49	8.28	9.53
Braintree ... Arr.	8.35	10.20	11. 9	1. 0	2. 5	2.35	3.23	4. 5	4. 5	5.45	6.45	7.55	8.34	10. 2

				WO	SO		WO				SO	SO		SO	WO	WSO
	a.m.	a.m.	a.m.	p.m.	p.m.	p.m.	p.m.	p.m.	p.m.	p.m.	p.m.	p.m.	p.m.	p.m.	p.m.	p.m.
Braintree ... Dept.	8.50	10.30	11.30	1. 0	1.15	2.45	3. 0	3.45	4.15	6. 0	7.15	8. 0	8.30	9. 0	9. 0	10. 5
RIFLE HILL	8.54	10.36	11.36	1. 6	1.21	2.51	3. 5	3.51	4.21	6. 6	7.21	8. 5	8.36	9. 5	9. 6	10.11
SANATORIUM	9. 1	10.43	11.43	1.13	1.29	2.58	3.10	3.58	4.28	6.13	7 28	8.10	8.43	9.10	9.13	10.18
WHITE NOTLEY (X Keys)	9. 6	10.49	11.49	1.20	1.37	3. 3	3.16	4. 4	4.32	6.19	7.32	8.16	8.49	9.16	9.19	10.24
FAULKBOURNE (P. Office)	9. 9	—	11.53	1.24	1.41	—	3.20	4. 8	4.36	6.23	—	—	8.53	—	9.23	10.28
VICTORIA, Power's Hall End	9.15	—	11.59	1.30	1.47	—	3.26	4.14	4.42	6.29	—	—	8.59	—	9.29	10.34
CHIPPING HILL	9.17	—	12. 2	1.32	1.50	—	3.28	4.16	4.44	6.31	—	—	9. 1	—	9.31	10.36
WITHAM	9.20	—	12. 5	1.35	1.53	—	3.30	4.20	4.47	6.35	—	—	9. 5	—	9.35	10.40
Hatfield Peverel Arr.	9.30	—	12.15	--	2. 3	—	—	4.30	5. 7	6.45	—	—	9.15	—	9.45	10.50

SO.—Saturdays Only. WSO.—Wednesdays and Saturdays Only. WO.—Wednesdays Only. NW.—Not Wednesdays.

SUNDAYS ONLY.

	a.m.	p.m.	p.m.	p.m.	p.m.	p.m.	p.m.	p.m.	p.m.	p.m	p.m.
Hatfield Peverel Dept.	9.30	12.15	—	1.50	—	—	—	5.30	7.15	8. 0	9.10
WITHAM	9.45	12.25	—	2. 0	—	—	—	5.35	7.20	8.10	9.15
CHIPPING HILL	9.48	12.28	—	2. 4	—	—	—	5.39	7.24	8.13	9.19
VICTORIA, Power's Hall End	9.50	12.30	—	2. 6	—	—	—	5.41	7.26	8.15	9.25
FAULKBOURNE (P. Office)	9.57	12.37	—	2.12	—	—	—	5.47	7.32	8.22	9.27
WHITE NOTLEY (X Keys)	10. 1	12.41	—	2.16	2.37	—	4.40	5.51	7.36	8.26	9.31
SANATORIUM	10. 7	12.47	1.35	2.22	2.43	3.20	5. 0	5.57	7.42	8.32	9.37
RIFLE HILL	10.14	12.54	1.42	2.29	2.50	3.27	5. 7	6. 4	7.49	8.39	9.44
Braintree ... Arr.	10.20	1. 0	1.48	2.35	2.56	3.33	5.15	6.10	7.56	8.45	9.50

	a.m.	p.m.	p.m.	p.m.	p.m.	p.m.	p.m.	p.m.	p.m.	p.m.	p.m.
Braintree ... Dept.	10.45	1.15	2.15	3. 0	3.45	4.15	5.20	6.30	8.15	9. 0	10. 0
RIFLE HILL	10.51	1.21	2.21	3. 5	3.51	4.21	5.25	6.36	8.21	9. 6	10. 6
SANATORIUM	10.58	1.29	2.28	3.12	3.58	4.28	5.32	6.43	8.26	9.13	10.13
WHITE NOTLEY (X Keys)	11. 4	—	2.34	—	4. 4	4.32	5.38	6.49	8.35	9.19	10.19
FAULKBOURNE (P. Office)	11. 8	—	—	—	4. 8	—	5.42	6.53	8.39	9.23	10.23
VICTORIA, Power's Hall End	11.14	—	—	—	4.14	—	5.49	6.59	8.45	9.29	10.29
CHIPPING HILL	11.16	—	—	—	4.16	—	5.51	7. 1	8.47	9.31	10.31
WITHAM	11.20	—	—	—	4.20	—	5.53	7. 5	8.52	9.35	10.35
Hatfield Peverel Arr.	11.30	—	—	—	4.30	—	6. 3	7.15	9. 2	9.45	10.45

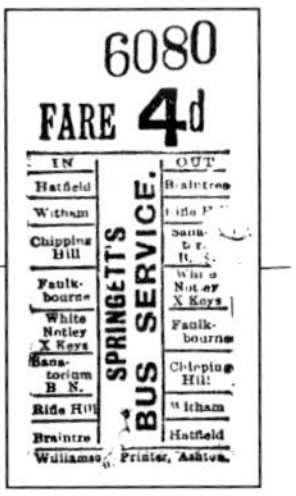

1. No responsibility will be held for any inconvenience caused by any departure from the above times.
2. No responsibility beyond ticket value.
3. No responsibility held for injury to passengers entering or leaving the 'Bus whilst it is in motion.
4. The right held to withdraw any Service without previous notice.

The Shearcroft Printing Co., Ltd., Market Place, Braintree.

Fig. 13 Springett's timetable and ticket.

which in 1935 Moore's offered £650 (approx. purchasing power of £22,000 in 1996) and reference to Appendix III will show that Moore's application for the transfer of the licence appeared in N&P No. 240 of 27th January 1936. An objection was received from Hicks Bros. and following a traffic court hearing the transfer was refused — with confirmation being given in N&P No. 248 of 23rd March 1936. In the meantime, Hicks applied to the Traffic Commissioners to operate a new service along the same route, to which the London North Eastern Railway Company then objected! A licence to run this service was also refused.

It is interesting that the Chairman of the Traffic Commissioners said at the hearing that to refuse transfer to Moore's could leave Springett without a willing buyer, also refusal might find Hicks objecting to a payment of £650 if they had intentions of buying the business later on. However, in spite of these considerations the decision was still taken to refuse a transfer. An account of the days hearing at the Traffic Court in Cambridge appeared in the Essex Weekly News dated 6th March 1936 as shown opposite:-

Telephone—KELVEDON 6.

Moore Bros. (Kelvedon) Ltd., Kelvedon.

Motor Contractors and Saloon Bus Proprietors.

CHELMSFORD TO COLCHESTER

AND

CHELMSFORD TO MALDON SERVICES.

Traffic Notice.

EXTENSION TO COVAL LANE.

Passengers holding Weekly Season or Return Tickets or Passengers intending to take such Tickets are allowed to use this extension between Chelmsford Station and Coval Lane; other Passengers are requested to alight at Chelmsford Station on the inward Journeys and to board the coaches at Market Road on the outward Journeys.

Chas. E. Leach, Printer, Tiptree.

Fig. 14 Passengers can now use Coval Lane. c1934

AMALGAMATION OF BUS SERVICES

HATFIELD - BRAINTREE ROUTE PROPOSAL

The proposed agreement for the transfer of a bus service between Hatfield Peverel and Braintree to be run in conjuction with the Chelmsford-Colchester services, for payment of £650, was before the Traffic Commissioners at Cambridge on Wednesday.

Messrs. Moore Bros. (Kelvedon), Ltd., applied for the transfer of the business of Messrs. H. Springett and Son, and it was opposed by Messrs. Hicks Bros., Ltd., of Braintree.

Mr. J. Nolan explained that Messrs. Moore operated services between Chelmsford and Colchester, via Witham and Braintree and via Coggeshall and Marks Tey. The proposal was for a daily service with two vehicles instead of one, operating on Wednesdays, Saturdays and Sundays. Co-ordination with two services now operated by Messrs. Moore was to follow the transfer. The extension of Black Notley Sanatorium was a reason for giving a better service.

Horace Bramley Moore said Mr. Springett informed him that the business was for sale, and he offered £650 for it. At that time he was not aware that the receipts per car mile, as the Chairman pointed out, were as low as 3½d., but he believed he could make the service pay.

The Chairman : There is a good margin for brains here with a loss of 2½d. per car mile.

Mr. Springett said his father was 70 years of age, but if the business were not transferred to Messrs. Moore he would like to continue it.

The Chairman : It would clear up a point right away if we knew whether Mr. Hicks is willing to pay £650 for the business of Mr. Springett.

Mr. Thesiger, counsel for Messrs. Hicks : I can answer that shortly - "Yes". It would not involve us in unremunerative running, and perhaps we are justified in paying more than it is worth.

Mr. Moore said that if the transfer were not granted the £100 deposit on the purchase, held by a solicitor, would be returned to him.

The Chairman : The effect of refusing the transfer to Mr. Moore might be to leave Mr. Springett without a willing purchaser. If the transfer is refused to Mr. Moore, Mr. Hicks might find all sorts of objections to paying £650 for the business. The Commissioners will not come to a decision at the present time because to do so may prejudice Mr. Springett's position considerably.

Accordingly the application was adjourned.

On 30th March 1936 an application by Hicks for the transfer of Springett's licence was given in N&P No. 249 and the grant for this appeared in N&P No. 256 dated 18th May 1936. This gives a good illustration of the way in which an operator could attempt to alter the course of events if it was believed the results were detrimental to their business interests. Later in 1936 Hicks terminal point in Hatfield Peverel changed from The Crown to Church Lane, which was a short distance up The Street towards Colchester. In 1949 Hicks applied to cutback the service to Witham High Street which was granted.

Mr Springett's garage became derelict over the years and nothing remains today apart from the overgrown site.

In common with many bus companies Moore Bros. used the suffix from their licence number to identify each of their service routes. These suffixes were to appear on time- and fare tables from about 1937 until the end of 1963, although this was not a requirement of the Traffic Commissioners.

As early as 1924 Moore's had introduced a form of route numbering; the sequence following the order in which services came into operation. These numbers can be clearly seen on a timetable published about this time, see Fig. 15. However some of these only served a short life since they were dropped in favour of those initiated by Colchester Corporation in 1926 to identify services using the new bus station in St. John's Street. A very large timetable board was erected by the Corporation at the front of the bus park which showed the services operating from each stand, against which a service number was given. These numbers were only used by Moore's for a short time and then not widely.

A cross index of all service numbers used by Moore's is included in the Network Growth Chart Appendix I. Unlike some bus companies Moore's service numbers never appeared as part of the vehicle destination blind.

Commencing April 28th, until further notice.

Telephone:—Kelvedon, P.O. 6

MOORE BROS.,

KELVEDON,

Motor 'Bus Services.

Service 1.

Kelvedon Station and Coggeshall.

	a.m.		a.m.
Leave Kelvedon	7.45	Leave Coggeshall	8.40
" "	10.18	" "	11.0
	p.m.		p.m.
" "	12.20	" "	1.10
" "	4.59	" "	7.0

Service 2.

Kelvedon, Coggeshall, Marks Tey, Colchester.

Daily, except Thursdays and Sundays.

	W.O.		S.O.	S.O.	N.S.	S.O.	S.O.		S.O.
Kelvedon	9·30	9·55	11·10	1·20	1·45			5·0	
Coggeshall	9·45	10·15	11·30	1·40	2·0	1·45	5·0	5·30	8·15
Marks Tey	10·5	10·35	11·50	2·0	2·20	2·5	5·20	5·50	8·35
Colchester	10·35	11·0	12·15	2·25	2·50	2·30	5·45	6·15	9·0

		S.O.	S.O.		W.F.S.	S.O.		S.O.	S.O.
Colchester	12·15	12·45	3·30	4·0	6·0	6·45	7·0	9·30	9·45
Marks Tey	12·40	1·5	3·55	4·25	6·25	7·10	7·25	9·55	10·10
Coggeshall	1·0	1·30	4·15	4·45	6·45	7·30	7·45	10·15	10·30
Kelvedon	1·20			5·5	7·5	7·50	8·0*	10·35	10·50

*On Saturday, stops at Coggeshall.

Service 3.

Kelvedon, Tiptree, Birch, Colchester.

Daily, except Thursdays and Sundays.

	S.O.	*	S.O.			S.O.	S.O.
Kelvedon ...	9·30	10·0	1·30			5·15	
Tiptree (Ship)	9·45	10·15		1·30			
Tiptree (Co-op)	10·0	10·30	1·45	1·45	5·15	5·30	8·0
Birch	10·20	10·50	2·5	2·5	5·35	5·50	8·20
Colchester ...	10·45	11·15	2·35	2·30	6·0	6·20	8·45

*Does not leave the Ship on Mondays, Tuesdays & Saturdays.

		S.O.		S.O.		S.O.
Colchester ...	12·15	3·30	4·0	6·45	7·0	9·30
Birch	12·40	3·55	4·25	7·10	7·25	9·55
Tiptree (Co-op)	1·0	4·15	4·45	7·30	7·45	10·15
Tiptree (Ship)	1·10	4·25	4·55	7·40		
Kelvedon ...		4·40	5·5		8·5	10·35

Service 4.

Kelvedon, Coggeshall, Earls Colne, Halstead, Braintree.

Wednesday only.

Kelvedon ...	10·0		Braintree ...	12·30	4·15
Coggeshall ...	10·15		Halstead ...	1·0	4·45
Earls Colne	10·35		Earls Colne		5·5
Halstead ...	11·0	1·15	Coggeshall ...		5·25
Braintree ...	11·30	1·45	Kelvedon ...		5·40

Service 5.

Braintree, Coggeshall, Marks Tey, Colchester.

Wednesday only.

Colchester	10·45	Braintree	3·45
Marks Tey	11·10	Coggeshall	4·15
Coggeshall	11·30	Marks Tey	4·35
Braintree	12·0	Colchester	5·0

Service 6.

Kelvedon, Marks Tey, Colchester.

Fridays only.

	F.O. A.M.	F.O. P.M.	S.O. P.M.
Kelvedon	9·40		8·15
Little London	9·45		8·20
Marks Tey (Trowel and Hammer) ..	9·55	1·45	8·30
Marks Tey (School)	10·0	1·50	8·35
Copford	10·5	1·55	8·40
Colchester	10·25	2·15	9·0

	S.O.	F.O.	F.O.
Colchester	11·0	1·0	3·45
Copford	11·20	1·20	4·5
Marks Tey (School)	11·25	1·25	4·10
Marks Tey (Trowel and Hammer) ...	11·30	1·30	4·15
Little London	11·40		4·25
Kelvedon	11·45		4·30

Fig. 15 Timetable c1924.

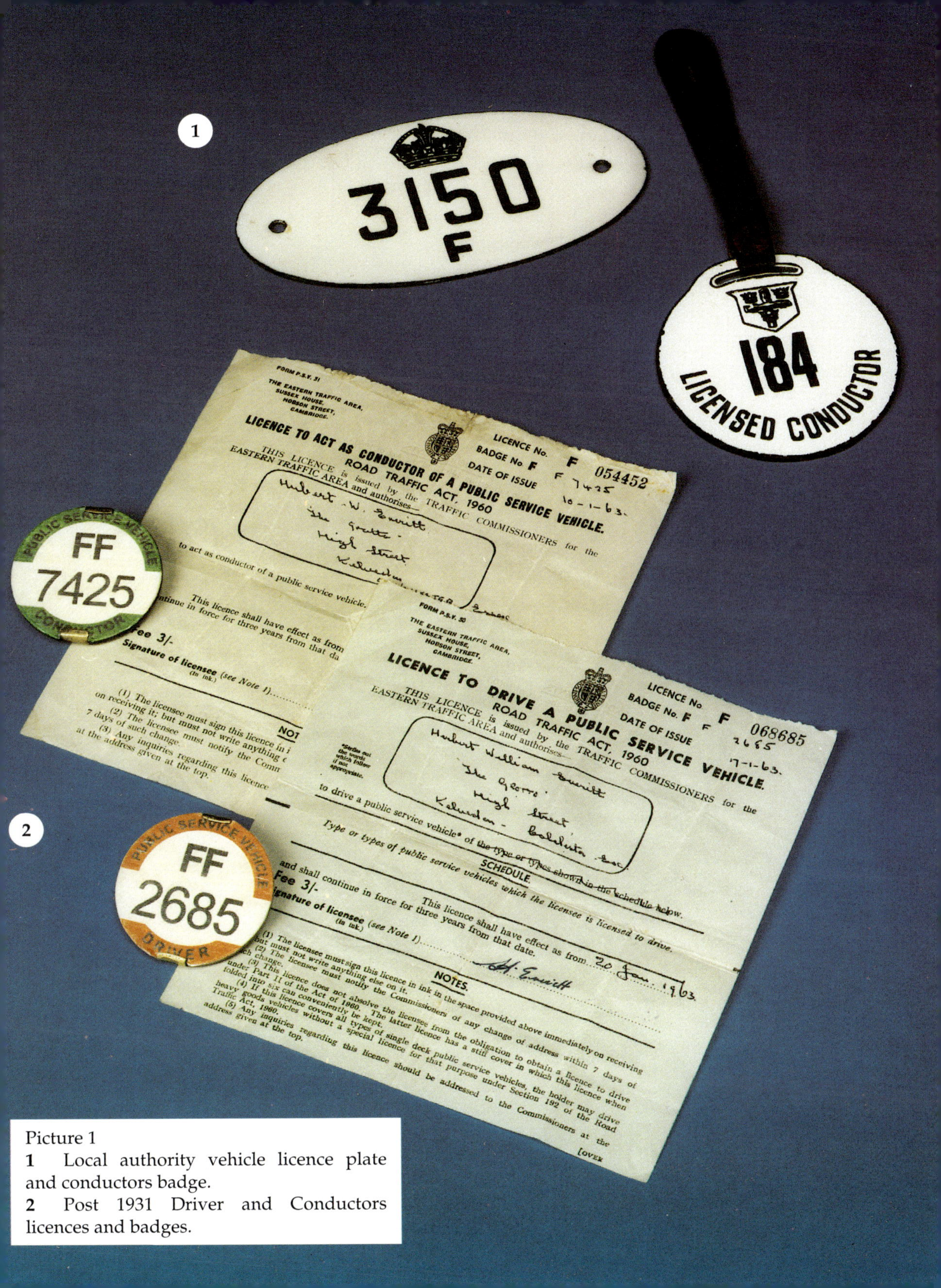

Picture 1

1 Local authority vehicle licence plate and conductors badge.

2 Post 1931 Driver and Conductors licences and badges.

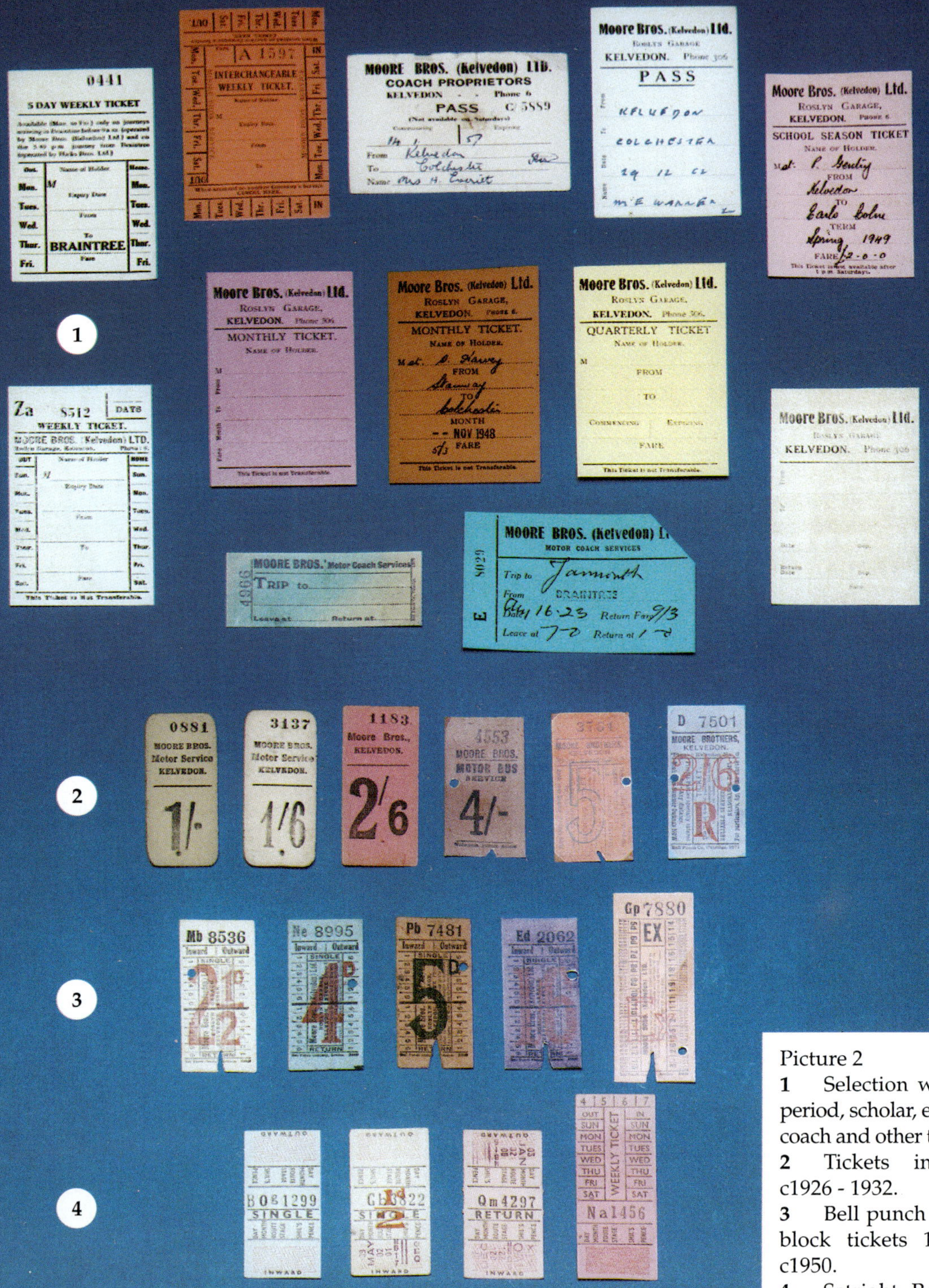

Picture 2

1 Selection weekly, period, scholar, express coach and other tickets.

2 Tickets in use c1926 - 1932.

3 Bell punch and B block tickets 1932 - c1950.

4 Setright Register tickets c1950 - 1963.

Picture 3
The Conductors Lot!

Picture 4
Company badges.

Picture 5
Company uniform.

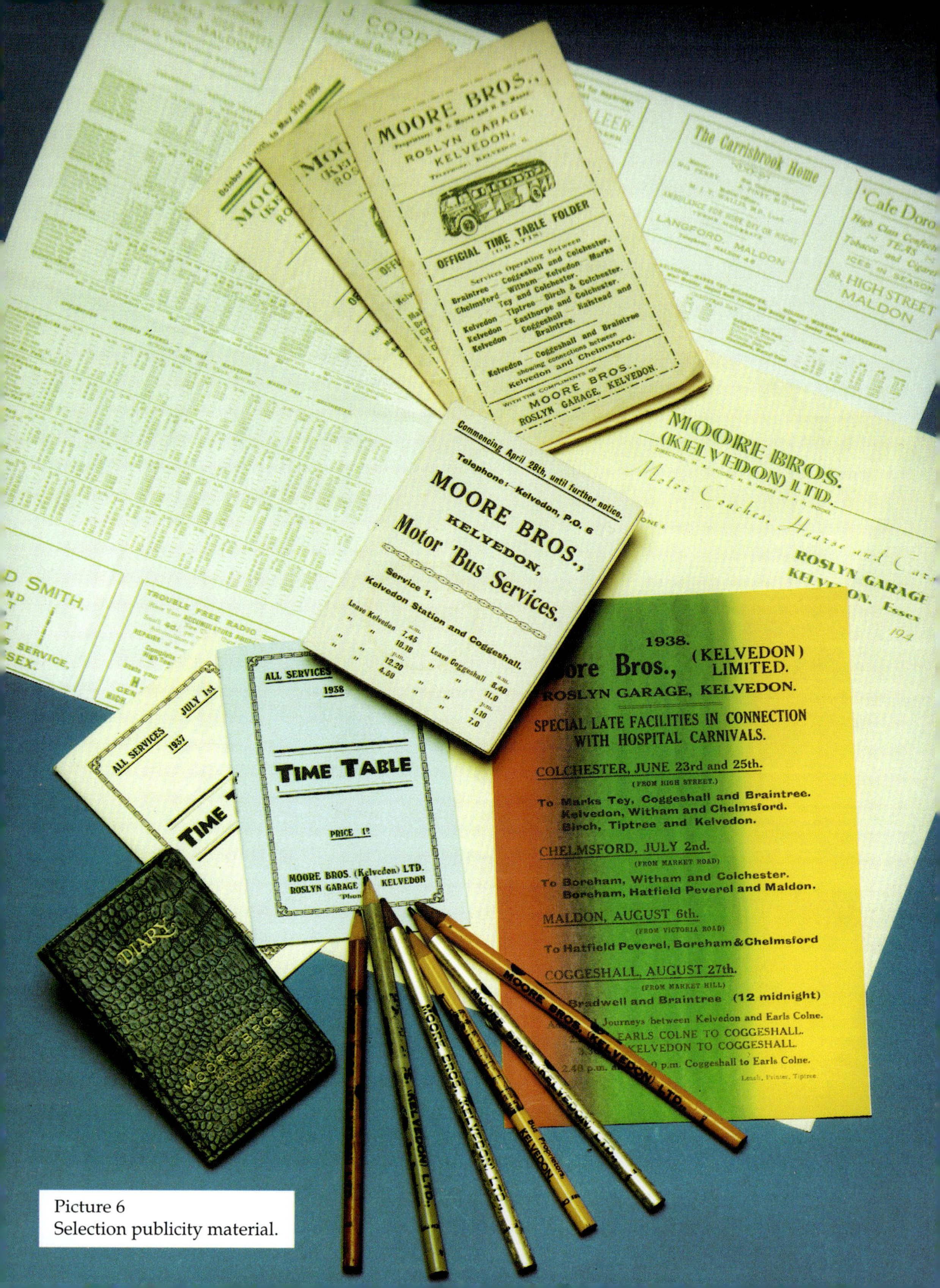

Picture 6
Selection publicity material.

Picture 7
Some of the chassis names in the fleet.

Picture 8
Moore's family and ex-staff, commemorative run 1993.

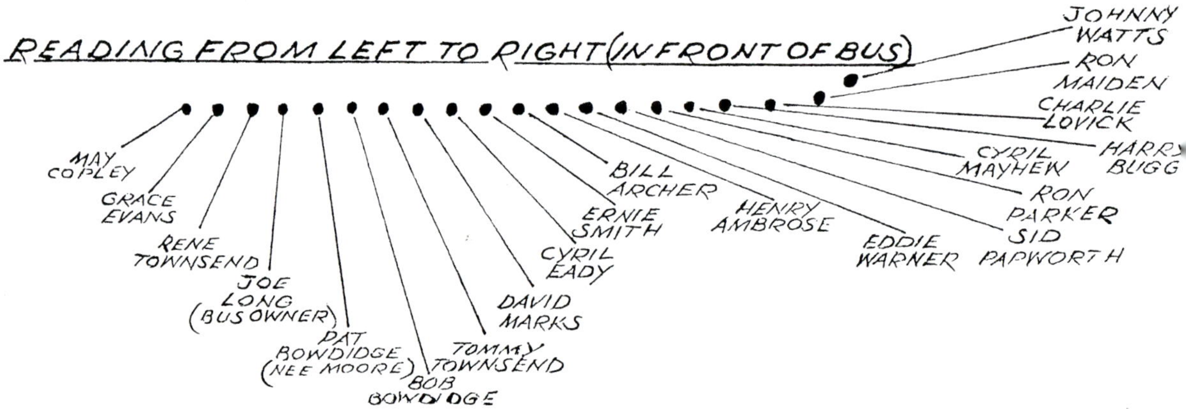

Chapter 8

The Prosperous Thirties

The 1930's was an important decade in the life of Moore Bros. as it brought great opportunities for development and prosperity. Regulation had stabilised the bus industry and as Moore's had received most of the 'Road Service Licences' they applied for their prospects looked good. Initiative, attention to detail and a business like approach to all things had been the hallmark of Charles and Edward's leadership. They had brought in timetables, business cards, and had made use of advertising in local newspapers and other publications. Charles' two sons, William and Horace, now owners of the business, were to follow in their father's footsteps. They expanded the motor bus fleet in the 1920's and at the same time introduced a smart green and gold livery to improve identity and create a public image. Attention was now to be given to bus crews so that their appearance would complement the vehicles in which they served.

It is very difficult to be sure when uniforms were first introduced. Undoubtedly a form of dress was expected from staff during the horse-drawn era but there is no evidence of a regular uniform. Special occasions and funerals would demand some degree of decorum, but as the owners would generally be present at such events formality would be looked after by them. Typical garments of the period can be seen in Pic. 5 in Chapter 3 where a long smock, top hat and high boots are in evidence, as well as a frock coat and bowler hat.

Reference to the driver of the steam bus in Pic. 21 in Chapter 5 will show that the white dust coat, so typical of the 1930's, had arrived, together with a stiffened driver's cap. However, this apparel was reserved for private parties and outings and not stage carriage work. The larger companies did provide a full uniform as can be seen in Pic. 20 in Chapter 5, but the private independent operator generally made no such provision.

The once familiar softened peaked cap had become available to all crew by 1924, with a white top cover for summertime use, but no cap

33 Doris Fincham wearing pre-war style overcoat.

badge. It was to be 1927 before the script 'Moore's' badge had become available. The original badge which is shown in Colour Pic. 4 was hand cut from solid brass and cost 1s.6d. My memory tells me my father had it made in a shop in Eld Lane, Colchester. William Moore seeing the badge asked him to get a further six made — at the firm's expense!! Some later batches were machine-stamped from soft metal and then brass plated. The attractive rounded edges of the letters in the original had been lost with this version. The dark blue serge overcoat, or greatcoat, was issued from about 1930 onwards. These had leather cuffs, elbow and pocket protectors. Later on, 'Moore Bros.' in script form was embroidered on the collar. This little feature gave a uniqueness to the company's coat — a picture of one of these appears in Pic. 33. Unfortunately, it was not repeated on coats produced after World War II .

We have already mentioned use of the summer white dust coat which generally was available only for excursion and private hire work. During the early 1930's these had become quite attractive with green collars and cuffs and 'Moore Bros.' embroidered in white on the collar. One essential item issued to all staff during World War II was the civilian version of the 'tin hat'. This was suitably painted in green and cream with 'MB' stencilled on the front.

The script 'Moore's' badge was replaced about 1950 with a green enamel version, and later still another version became available which had the advantage of being suitable for use on both the cap and collars of the post World War II greatcoats and dust coats. These can be seen in Colour Pic. 4.

The first post World War II dust coat although still full length was more practical in colour and was issued to all crews for summer use. It was quite a bright green cloth with beige cuffs and collar, on which 'Moore Bros.' was embroidered in light green cotton. Later on, this was replaced

34 J type Thornycroft now fitted pneumatic tyres. Rebodied 32 seat bus Strachan and Brown c1929.

with a summer jacket which was mainly khaki in colour with green cuffs and collar, 'Moore Bros.' again being embroidered on the collar. Examples of these uniform items can be seen in Colour Pic. 5.

We saw in the previous chapter how at the start of the 1920's Moore's had invested in Thornycroft double-deckers for use on more heavily populated routes, whereas the Model T Fords, because of their small size and capacity, were suited to sparsely populated routes and narrow country roads. By the end of the decade the open-top Thornycrofts were being dispensed with in favour of single-deck Thornycrofts, see Pic. 34. Although these had reduced seating capacity, the need for high capacity vehicles no longer applied because of increased frequency in services. All of this was in the best interests of the travelling public.

With the arrival of the 1930 Road Traffic Act speeds of up to 30 m.p.h. were now allowed on PSV buses. Like many independent operators Moore's turned their attentions to the Gilford make of chassis and those fitted with Petty coachwork were quite luxurious for their time, see Pic. 35. Seats were well sprung, with handrails and cut-glass ashtrays fitted to the seat backs, and curtains at the windows, all adding to passenger comfort. Interior lighting was to a high standard with attractive

35 Petty bodied Gilford. Market Hill, Coggeshall c1930. G. Mayhew driver, and J. Pitchford.

35a Gilford Wycombe body. GMC alongside. Colchester c1931.

cut-glass bowls fitted to the lamps. With Gruss air springs as part of the front suspension, passengers could be sure of a comfortable ride. The Gilford Motor Company went into receivership in 1935, but Moore Bros. had made no further purchases of this make since 1932.

As we enter the 1930's, 32 seater saloons were in use on most of the main service routes and it had become company policy for a crew to be allocated to a specific vehicle. The more senior drivers were given new vehicles. Whilst this policy gave a sense of pride of ownership to the crew, it did create a problem in that as long as a bus was on service so were the crew. This frequently meant a very long working day. This policy remained unchanged until the arrival of double-deckers when shift working was introduced.

Meanwhile 1932 became a major milestone in the history of the firm. On 1st October it was registered at Companies House with a new name Moore Brothers (Kelvedon) Limited under No. 268956, as a private limited company with its Registered Offices at Roslyn Garage, High Street, Kelvedon. The two directors were William C. Moore and Horace B. Moore. The Company was formed to take over and carry on the business of omnibus, and taxicab proprietors and car hire owners. Capital authorised £8,000 (in 1996 figures approx. £250,000), issued £7,500. Debentures were authorised and issued to value £5,000 on 22nd November 1932. Share ownership was limited to close relatives. Interestingly, the Moore brothers retained ownership of the land and property and this was not transferred to the Company until 1940.

With the formation of a limited company on 1st October 1932 the Deed of Partnership which William and Horace signed in November 1928 was no longer relevant and therefore was dissolved at this time. However, this left a problem since in the event of a dispute between the brothers, a conflict of interest could arise between the limited company and private ownership of the property. Clearly the company required premises from which to operate. A Declaration of Trust was signed on the 7th February 1933. This made provision for each brother to have a right to purchase the other brother's equal share in the property, should there be a wish to sell that share. As we shall see in the next chapter an opportunity to exercise this right arose in 1940.

Following the sale of the company to Eastern National in December 1962, Moore Brothers (Kelvedon) Limited was dissolved on 1st February 1965. Regrettably, the file held by Companies House was destroyed about 1975.

'Phone 6.

Moore Bros. Motor Proprietors, Kelvedon

1932 SEASON (ending Sept. 30th). ESSEX.

EXCURSIONS & TOURS from

BRAINTREE, COGGESHALL, KELVEDON & TIPTREE

CLACTON-ON-SEA

From BRAINTREE—every Sunday, Tuesday and Thursday,
Dept. 8.30 a.m. Rtn. 6.30 p.m. **4/6** Return.

" COGGESHALL—every Sunday, Tuesday and Thursday,
Dpt. 8.50 a.m. Rtn. 6.30 p.m. **3/6** Return.

" KELVEDON—every Sunday and Thursday,
Dpt. 8·30 a.m. Rtn. 6·30 p.m. **3/6** Return.

" TIPTREE—every Sunday and Thursday,
Dpt. 8·45 a.m. Rtn. 6.30 p.m. **3/6** Return.

MERSEA ISLAND

From BRAINTREE, dpt. 2 p.m. **2/6** rtn. KELVEDON, dpt. 2.30, **1/6** rtn.
COGGESHALL " 2.20 **1/6** " TIPTREE " 2·45, **1/6** "
Every Sunday and Thursday. Leaving Mersea Island 7 p.m.

CIRCULAR EVENING TRIPS

Every Sunday from BRAINTREE to
BULMER, HEYBRIDGE or FINCHINGFIELD, dpt. 6.15 p.m. Fare **1/6**

LOWESTOFT & YARMOUTH

Every Saturday, from July 30th to Aug 27th inclusive.
(Weekly Period Returns only)

From BRAINTREE, dpt. 7 a.m. Leaving Yarmouth 1 p.m.
BRADWELL, " 7.12 " " Lowestoft 1·30 "
COGGESHALL " 7.25 "

Return Fares:—to Lowestoft **15/-** To Yarmouth **16/-**

Additional tours to Yarmouth, Lowestoft, Southend-on-Sea, Felixstowe, Whipsnade Zoo, Clacton-on-Sea and Mersea Island
Will be operated during the season.

For Further Particulars please apply to our Agents, where seats should be booked, or at our Office—
ROSLYN GARAGE, KELVEDON.

Fig. 16 Excursions and tours 1932.

We have seen how charabancs had been purchased in the 1920's to satisfy the increasing demand for day trips in summer to the seaside resorts. By the 1930's the popularity of the charabanc was waning in preference to the saloon, where passengers could still enjoy open skies by way of the folding or sliding sunshine roof.

Day excursions and evening circular trips had really got into their stride by the summer season of 1932 as reference to the programme in Fig. 16 will show. Perhaps surprising is the strength in which these tours were operated from Braintree, the domain of Hicks Bros. The importance that Moore's attached to this aspect of the business can be gauged by the appointments of Rudkin in Braintree, Humphrey in Coggeshall, and Moore in Tiptree as booking agents. A display board outside Eva Moore's shop advertising a trip to Clacton for 3s.6d. return can be seen in Pic. 36. The photo was taken about 1935.

The private hire business was developing also and perhaps the best example of this is the Co-op outing from Coggeshall in June 1937, see Pic. 37. At the other end of the scale we find a bus being provided to the British Legion for taking a party to Hyde Park, London, where a review of some 80,000 ex-service men and women was to take place by his Majesty King George VI. A picture showing the programme front cover of this special event together with Moore's invoice for hire of the Albion Saloon is in Fig. 17. Hiring coaches to the Women's Institute and many other private organisations for day outings had become quite common during the 1930's. I well remember my father taking a coach party of ladies to the Vale of Evesham in Worcestershire to

36 Moore's agent in Tiptree, Eva Moore. Display board reads 'Moore's Bus Service, August Bank Holiday Special Excursion, to Clacton from Factory Corner 8.45 am. Return from Clacton 6.30 pm. Return fare 3/6. Please Book Now.'

see the apple blossom and being rather disgruntled that the firm had accepted such a long journey for a day's outing. It was the last straw when the lady in charge of the party said; 'Driver — can we go home a different way please?' His reply was; 'Certainly, Madam' — and he promptly drove home by the same route, but now on the other side of the road!!

The period also saw the smaller village school closing down, or the older children being moved to another school in the vicinity. Certain schools provided classes of specialised subjects in which boys and girls from other schools would need to attend. All this necessitated the use of buses which would be provided under contract to education authorities. Typical examples of those provided by Moore's were taking children from Easthorpe and Copford Green to Birch, Fairstead and Terling to Witham, and Pattiswick to Braintree.

More children were attending the high schools in Chelmsford, Colchester, Braintree, Halstead and Maldon requiring additional seating capacity on scheduled services during school travelling times. This was

37 Coggeshall Co-op outing 1937 involving 6 Moore's coaches plus 2 others - possibly Blackwell.

REVIEW

OF

Ex-Service Men and Women

BY

His Majesty the King

accompanied by the Queen.

AT

HYDE PARK

On Sunday, June 27th, 1937

at 2.30 p.m.

Public Admission to Hyde Park will only be by Stanhope Gate and the gates leading from Kensington Gardens.

Souvenir

PROGRAMME

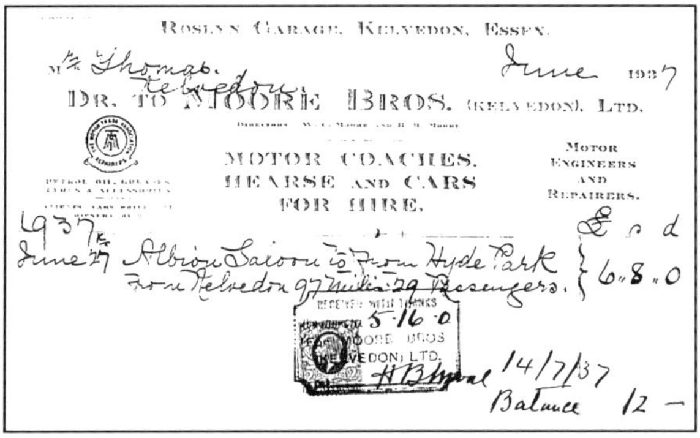

ROSLYN GARAGE, KELVEDON, ESSEX.

Mr Thomas.
Kelvedon — June 1937

DR. TO MOORE BROS. (KELVEDON), LTD.

MOTOR COACHES,
HEARSE AND CARS
FOR HIRE.

MOTOR
ENGINEERS
AND
REPAIRERS.

		£ s d
1937 June 27	Albion Saloon to From Hyde Park From Kelvedon 97 Miles 29 Passengers.	6. 8. 0

RECEIVED WITH THANKS 5-16-0
FOR MOORE BROS (KELVEDON) LTD.
H B Moore 14/7/37

Balance 12 –

Fig. 17 Outing to Hyde Park, London. 1937.

often provided by buses following their works service duties — the reverse taking place in the evenings. School season tickets then became available, also tickets were interchangeable with Eastern National and Hicks Bros. on certain services. The growth of employment in factories was continuing and to support that trend works services were introduced which would take workers from Tiptree, Coggeshall, Kelvedon, and along those routes directly to the factory gates of Crittall, Marconi and Hoffmann.

With more and more places being reached by bus a seaside trip was a great attraction during the summer months, particularly when children were on holiday. With this in mind Moore's introduced in 1933 a stage-carriage service on Wednesdays and Saturdays from Braintree through Coggeshall, Kelvedon and Tiptree to Mersea Island. At that time a further application was made to operate a similar service from Braintree but via Marks Tey, Stanway, Lexden, Shrub End, Old Heath, Abberton and Peldon. Not surprisingly there was strong opposition to this coming from, A. W. Berry, Eastern National, The Mersea Colchester and District Transport and Bus Co. Ltd., and P. H. Underwood. The licence was not granted. Later on a service was opened up from Earls Colne through

Coggeshall, Kelvedon, Rivenhall End and Wickham Bishops to Maldon on Wednesdays and Sundays. Both services were to operate during the summer months only. In reality the latter service was an extension to the daily all year round service between Kelvedon and Earls Colne which had been started in 1933. The application to the Traffic Commissioners provided for fare stages along the whole route from Earls Colne to Maldon and according to the Notices and Proceedings a licence was granted on this basis. The service started on 1st July 1937 but the timetable from that date shows that between The White Hart, which was near to The Angel corner in Kelvedon, and Victoria Road Maldon, no passengers could be picked up for destinations between these two points. Whether the Traffic Commissioners subsequently imposed this condition, or whether it was inadvertently omitted from the N&P's I have not yet established. Obviously this was a serious restriction of Moore's aims.

With the expansion of stage services and provision of other facilities there was a need to increase public awareness of them. Whilst the buses and bus parks displayed timetables and notices, more attractive publicity was needed. Coloured sheets were introduced to draw attention to later bus departures following carnivals and other special events. This system was also used to give information on additional services during national holiday periods. Then in 1932 for the first time a folded timetable appeared covering all services and specifically designed for issue to the travelling public. This was produced in the firm's colours of green and cream. It also carried a range of shop and business advertisements for the towns and villages that Moore's buses served. The revenue from advertising undoubtedly paid for the production of the booklet and allowed these to be available gratis . Later on 'Summer' and 'Winter' versions were introduced which provided all year round service details. These ceased in 1936 and were replaced by a 1937 version which was completely different, being devoid of all adverts and costing one penny a piece.

Complimentary annual pocket diaries, with Moore Bros. embossed on the front cover, became available, no doubt being reserved for worthy customers. Pencils in a wide range of colour finishes were produced and given out to lesser mortals. Examples of all of these publicity items are shown in Colour Pic. 6. World War II brought to an end the availability of this attractive publicity material which sadly was not to reappear in the post-war years. This was also to apply to certain services where the war years had forced their withdrawal.

We must now return to 1934 since this was an important year in the history of Moore Bros. for it saw the early introduction of standardisation into the vehicle fleet. Until then numerous chassis and body makes had been used with all the difficulties of maintenance and spares holdings that this brings. From now on and to the beginning of World War II Albion chassis and Duple coachwork would dominate the fleet — with Bedford fulfilling all the small capacity needs.

Albion started producing bus chassis in their factory at Scotstoun near Glasgow, Scotland, in 1925, although vehicle manufacture can be traced back to 1899. They had achieved a reputation for quality and reliability where their slogan 'Sure as the Sunrise' was enshrined in their radiator badge. These words were to be proven over the next few years on the roads of Essex, and well beyond. The Albions were more than a match for the Dennis Lancets on the main competing route between Chelmsford and Colchester being operated by Eastern National.

In 1931 Albion Motors introduced a new range of bus chassis which were named Valiant, Valkyrie and Victor. The first Albions to arrive at Moore Bros. were two Valiant saloons, BEV464 and BEV465, having 32 seats — this was in June 1934. A picture of BEV464 at the Duple bodyworks is shown in Pic. 38. This particular bus was allocated to my father and within a month of the picture being taken he drove it back to Duple to have accident repairs carried out as a vehicle had driven into the

38 Moore's first Albion Valiant at Duple bodyworks ready for delivery 1934.

39 AEC Regal, EV6337, new 1932, rebodied Duple 1937. Picture London c1952.

offside rear corner of the body. I accompanied him on the journey and well remember looking into the assembly bays at the Duple works and being amazed to see rows of Post Office Van bodies being built. As I was to discover later, not only were Duple main bodybuilders for the Post Office mail vans, but also vans for Post Office Telephones. These included the linesman's version where the top portion of the windscreen was tinted and slanted backwards at 45 degrees to allow inspection of the overhead telephone wires, of which there were many in those days.

In 1935 the nation celebrated the Silver Jubilee of King George V in a carnival-like atmosphere with processions parading through the streets of towns and villages alike. Kelvedon was no exception and Moore Bros. proudly entered their new Albion Valiant bus, BEV465, preceded by a carriers cart and two donkeys giving an impressive spectacle of contrasts between 1815 and 1935. However, some licence had been used in this portrayal since W.E.Moore was not the founder of the business nor had it been started by using two donkeys to haul the carriers cart. Even so, it didn't prevent the enjoyment of the occasion and the community was very proud to learn that this bus company had been part of village life for the past 120 years.

A third Valiant, CHK953, was delivered in 1935 still having 32 seats, but by this time Albion had dropped the 6-cylinder 85 Brake Horsepower (bhp) petrol engine, as fitted to the two previous Valiants, in favour of the larger 110 bhp unit which was being used in the Valkyrie. However, in 1936 the Valiant had ceased manufacture in favour of the Valkyrie type.

The year 1936 saw arrivals of both Valkyrie and Victor models, the Valkyrie having a longer wheelbase whereas the Victor had a shorter one compared to the Valiant. The Victor's engine was also smaller, only developing 65 bhp. Not surprisingly the Victor had seating capacity for only 31 passengers, but the Valkyrie had 35 seats. A further Victor was delivered in 1937 and this was the last Moore's Albion to have a petrol engine fitted. Interestingly at the time of the third Valiant delivery an AEC Regal, CHK952, was delivered carrying an almost identical Duple body. This was the second and last Regal to be taken into the fleet. The first Regal, EV6337, arrived in 1932 with a 32 seat Petty Bros. body, in common with those fitted to Gilford's being delivered about the same time. However, with the AEC being a better class of chassis the company must have decided it was able to serve a longer life and as an aid to this, a new Duple body was fitted in 1937, which with a swept waistline would help to keep abreast of fashion, see Pic. 39. Both AEC chassis were fitted with petrol engines, the Moore brothers having yet to be convinced of the benefits of oil engines which were becoming popular. However, this delay gave an advantage to the travelling public who could continue to enjoy the smooth running of the petrol engine.

Duple Bodies and Motors Ltd. had been producing coachwork for passenger carrying vehicles in London since 1919, first at their factory in Hornsey, then at Hendon where the bodies with which we are concerned were built, and finally bodybuilding was transferred from Hendon to Blackpool in Lancashire. Duple coachwork was also specified for all Bedfords delivered from 1934 onwards, although the very first Bedford to arrive was in 1932 which had a Petty body fitted.

Bedford bus manufacture started in August 1931, at that time only producing chassis suitable for 14 seat and 20 seat low capacity bodywork. These were produced at the Vauxhall Motor factory in Luton, where previously Chevrolet chassis had been manufactured. Chevrolet production ceased with the introduction of the Bedford range. Originally Chevrolet chassis had been assembled at Hendon employing components manufactured in Canada, but later production was transferred to Luton where most parts were then manufactured in the factory. Both Chevrolet and Vauxhall Motors were part of General Motors, USA. An extract from

Fig. 18 20 seater Duple bodied Bedford as displayed at Olympia 1931.

the Chevrolet and Bedford magazine of 1931 in Fig. 18 shows the introduction of the Bedford range of 14-seater WHB and 20-seater WLB bus and coach chassis. The bus illustrated would be very similar to the Duple-bodied Bedfords delivered to Moore's, the first of which arrived in 1934, BPU202 and BPU676. A distant view of BPU202 can be seen in Pic. 40 whilst standing at the Factory Corner, Tiptree, about 1935. A further WLB, CHK401, arrived in 1935 which was later to become Moore's breakdown truck. A picture of this in its converted state is shown in Pic. 41. At the end of 1935 Bedford introduced the WTB chassis which was designed to take 26-seater bus and coach bodies. This had a shorter bonnet and redesigned chromium plated radiator, improving the frontal appearance considerably. Moore's took delivery in 1937 of just one example of this, FEV806, which carried a 20-seater Duple body, as shown in Pic. 42.

The most significant arrival in 1937 was delivery of two of the three AEC Regent double-deckers, purchased second-hand from the City of Oxford; a third Regent arriving in 1938. The importance Moore's attached to the future use of double-deckers can be gauged from their application to the Traffic Commissioners in November 1937 for permission to use these vehicles on all their stage carriage routes. This was granted the following month, apart from the Braintree to West Mersea service where strangely approval was not given until July 1938. All double-deckers taken into the Moore's fleet, then and in the future, apart from two Leyland Titans, would be of the 'lowbridge' type having a sunken gangway on the upper deck. This allowed for a lower overall height of body which was desirable for a bus to pass under the railway bridge in

40 Factory Corner, Tiptree. Distant view Bedford BPU202. c1935.

Station Road, Kelvedon, and necessary for the one in Duke Street, Chelmsford. The ex-City of Oxford Regents had a most unusual seating arrangement on the upper deck with seats in rows of three and a sunken gangway along each side of the bus. A fourth Regent was added in 1941 which had been in the fleet of South Wales Transport Co. Ltd. All four

41 WLB Bedford, CHK401 now converted to breakdown truck.

42 WTB Bedford, delivered new 1937.

Regents were petrol engined. Examples of each vehicle type whilst with these operators can be seen in Pics. 43 and 44.

Two more Albion Valkyries, GNO267 and GNO268, were delivered in 1938 which were the last of this make to be received, but more significantly they were fitted with 5-cylinder Gardner diesel engines. The choice of Gardner was yet another step along the road of standardisation, for as we will see in later chapters this make of engine was to power most single and double-decker purchases up to the end of Moore's operations.

The last two 20-seater WTB Bedfords, GVW634 and GVW635, arrived during the same year, but these took on a distinctive new look having a rounded radiator grille with concealed radiator cap. This radiator front was the forerunner of the version fitted to the OB chassis which became so popular after the second World War. The 1938 version of the WTB chassis was also fitted with an improved engine developing 72 bhp instead of the earlier 64 bhp. The sleek looking Duple body with swept waistline all added to the modern appearance. A picture of GVW635 at the St. John's Street Bus Park, Colchester, can be seen in Pic. 45.

The 1930's had certainly been the heyday of the motor bus.

43 AEC Regent new to City of Oxford 1930. Example is of same batch delivered to Moore's in 1937/38.

44 AEC Regent WN4760 new to South Wales Transport Co. Ltd. 1932. Sister vehicle delivered to Moore's 1941.

44a Albion Valkyrie GNO268. Moore's first diesel engine.

45 Delivered in Dec 1938. Last pair Bedfords to Moore's before WW II.

Fig. 19 1939 timetable.

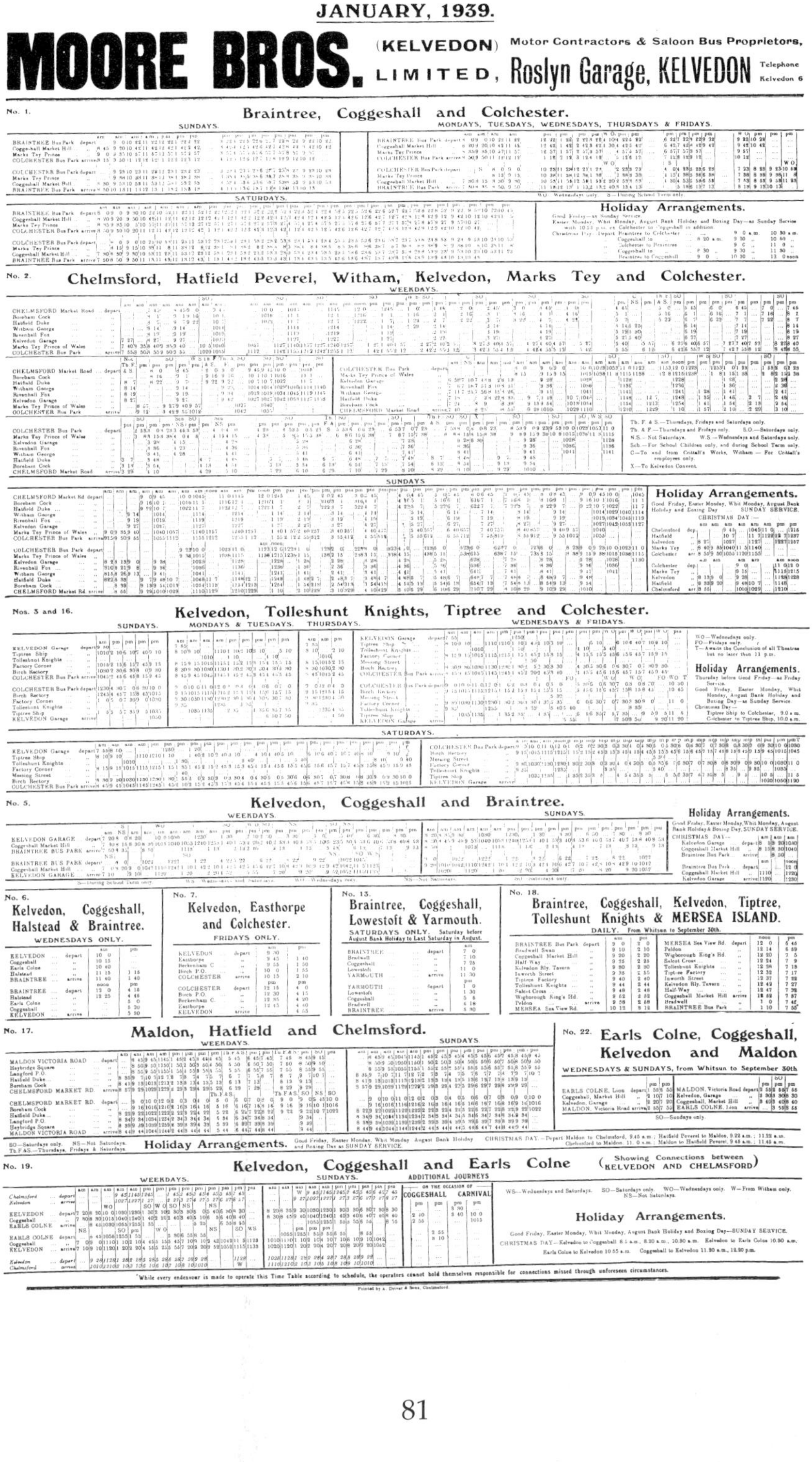

JANUARY, 1939.

MOORE BROS. (KELVEDON) LIMITED,

Motor Contractors & Saloon Bus Proprietors, Roslyn Garage, KELVEDON

Telephone Kelvedon 6

No. 1. Braintree, Coggeshall and Colchester.

SUNDAYS. MONDAYS, TUESDAYS, WEDNESDAYS, THURSDAYS & FRIDAYS. SATURDAYS.

WO—Wednesdays only. S—During School Term only.

Holiday Arrangements.

Good Friday—as Sunday Service.

Easter Monday, Whit Monday, August Bank Holiday and Boxing Day—as Sunday Service with 10.53 p.m. ex Colchester to Coggeshall in addition.

Christmas Day—Depart Braintree to Colchester ... ... 9 0 a.m. 10 30 a.m.
Coggeshall to ... 8 20 a.m. 9 20 .. 10 50 ..
Colchester to Braintree ... ... 9 0 .. 11 0 ..
Coggeshall to ... 8 30 .. 9 30 .. 11 30 ..
Braintree to Coggeshall ... 9 0 .. 10 30 .. 12 0 noon

No. 2. Chelmsford, Hatfield Peverel, Witham, Kelvedon, Marks Tey and Colchester.

WEEKDAYS.

Th. F. & S.—Thursdays, Fridays and Saturdays only.
Th. & F.—Thursdays and Fridays only. S.O.—Saturdays only.
N.S.—Not Saturdays. W.S.—Wednesdays and Saturdays only.
Sch.—For School Children only, and during School Term only.
C—To and from Crittall's Works, Witham—For Crittall's employees only.
X—To Kelvedon Convent.

SUNDAYS

Holiday Arrangements.

Good Friday, Easter Monday, Whit Monday, August Bank Holiday and Boxing Day ... SUNDAY SERVICE.

CHRISTMAS DAY:—

Nos. 3 and 16. Kelvedon, Tolleshunt Knights, Tiptree and Colchester.

SUNDAYS. MONDAYS & TUESDAYS. THURSDAYS. WEDNESDAYS & FRIDAYS. SATURDAYS.

WO—Wednesdays only.
FO—Fridays only.
T—Awaits the Conclusion of all Theatres but no later than 11 p.m.

Holiday Arrangements.

Thursday before Good Friday—as Friday Service.

Good Friday, Easter Monday, Whit Monday, August Bank Holiday and Boxing Day—as Sunday Service.

Christmas Day—
Tiptree Ship to Colchester, 9.0 a.m.
Colchester to Tiptree Ship, 10.0 a.m.

No. 5. Kelvedon, Coggeshall and Braintree.

WEEKDAYS. SUNDAYS.

Holiday Arrangements.

Good Friday, Easter Monday, Whit Monday, August Bank Holiday & Boxing Day, SUNDAY SERVICE.

CHRISTMAS DAY

No. 6. Kelvedon, Coggeshall, Halstead & Braintree.

WEDNESDAYS ONLY.

No. 7. Kelvedon, Easthorpe and Colchester.

FRIDAYS ONLY.

No. 13. Braintree, Coggeshall, Lowestoft & Yarmouth.

SATURDAYS ONLY. Saturday before August Bank Holiday to Last Saturday in August.

No. 18. Braintree, Coggeshall, Kelvedon, Tiptree, Tolleshunt Knights & MERSEA ISLAND.

DAILY. From Whitsun to September 30th.

No. 17. Maldon, Hatfield and Chelmsford.

WEEKDAYS. SUNDAYS.

SO—Saturdays only. NS—Not Saturdays. Th.F&S.—Thursdays, Fridays & Saturdays.

Holiday Arrangements. Good Friday, Easter Monday, Whit Monday, August Bank Holiday and Boxing Day as SUNDAY SERVICE.

CHRISTMAS DAY.—Depart Maldon to Chelmsford, 9.45 a.m.; Hatfield Peverel to Maldon, 9.22 a.m.; 11.22 a.m. Chelmsford to Maldon, 11. 0 a.m.; Maldon to Hatfield Peverel, 9.45 a.m.; 11.45 a.m.

No. 22. Earls Colne, Coggeshall, Kelvedon and Maldon

WEDNESDAYS & SUNDAYS, from Whitsun to September 30th

SO—Sundays only.

No. 19. Kelvedon, Coggeshall and Earls Colne (Showing Connections between KELVEDON AND CHELMSFORD)

WEEKDAYS. SUNDAYS. ADDITIONAL JOURNEYS — ON THE OCCASION OF COGGESHALL CARNIVAL

WS—Wednesdays and Saturdays. SO—Saturdays only. WO—Wednesdays only. W—From Witham only. NS—Not Saturdays.

Holiday Arrangements.

Good Friday, Easter Monday, Whit Monday, August Bank Holiday and Boxing Day—SUNDAY SERVICE.

CHRISTMAS DAY—Kelvedon to Coggeshall 8.5 a.m., 8.20 a.m., 10.30 a.m. Kelvedon to Earls Colne 10.30 a.m. Earls Colne to Kelvedon 10.55 a.m. Coggeshall to Kelvedon 11.20 a.m., 12.20 p.m.

While every endeavour is made to operate this Time Table according to schedule, the operators cannot hold themselves responsible for connections missed through unforeseen circumstances.

Printed by A. Driver & Sons, Chelmsford

Chapter 9

And Now It's Wartime

Before we enter into the detail of this period, one sad event that needs to be recorded was the death of William Moore on 23rd May 1940 at 76 years of age. From his earliest days William had been an active member of the local community. Before becoming a partner with his two brothers in the bus business of Moore Bros. in 1905, he had run an insurance agency from his home, 'Bridge House', at the bottom of Feering Hill. William had been a Parish Councillor for some years and was a member of the Temperance Society having been strictly brought up to avoid alcohol. His father and family had been lifelong abstainers and their children were expected to follow in their footsteps. William possessed a very kindly nature and I certainly can recall many acts of kindness to me as a child — including the one where he found me playing with a can of oil and insisted I continued to enjoy myself, much to my parents dismay! At the time of the Coronation of King George VI in 1937 both William and his wife, Rosanna, visited the schools in Kelvedon when each child was given a sixpenny piece. In the years preceding World War II William could often be seen standing outside his garage with 'Pinto', his Dalmatian dog, not far away. He always seemed to be wearing a straw boater, but I expect this was during summertime only.

You will recall in Chapter 8, William and his brother Horace continued to wholly own the firm's premises although a limited company had been formed to take over the bus business and its assets. Under the terms of the Declaration of Trust Horace could now exercise his right to purchase his late brother's share of the property, but he chose not to do so. Instead he abided by the wishes of his late brother's Will which provided for William's two sons Eric and Harold to inherit their father's share of the property. Agreement was then reached between the three parties that the property should be sold to Moore Bros. (Kelvedon) Ltd. for the sum of £2,290 (comparable to a purchasing power of approx. £77,000 in 1996). William's widow, Rosanna, who was now totally blind continued to live in Roslyn House until her death in 1949 at 82 years of age. William made a number of monetary bequests to members of his family, but also he left 500 ordinary shares in Moore Bros. (Kelvedon) Ltd. to his son Eric, thus ensuring his continued interest in the bus company.

Whilst on the subject of land it should be mentioned that in 1943 a strip of land three feet wide belonging to No. 6 Roslyn Cottages, was purchased from Ada Polley of Ormonde Lodge. This allowed the garage entrance to be widened which had always presented difficulties for drivers when negotiating a bus into the yard. To ease the problem of narrow access it had been a rule for sometime that buses were only permitted to enter and leave the premises from the Colchester direction and woe betide any driver caught by William or Horace attempting to do otherwise!! Purchase of this strip of land was fortuitous because later legislation permitted bus width to be increased to 8 foot and length to 30 foot and the extra land became very necessary.

We must now return to 1939 and the threat of war, the clouds of which had been gathering for some considerable time and when World War II finally broke out on 3rd September 1939 the British Government had many preparations in place. The last Notices and Proceedings of the Eastern Traffic Area to be issued prior to the outbreak of war was No. 411 on 28th August 1939. N&P's were not resumed until 1946 when the first post-war copy, No. 412, was issued on the 22nd July of that year. During September 1939 it was announced that the Traffic Commissioners were to be replaced by Regional Transport Commissioners, who for the Eastern Area would be based in Cambridge and occupy the same building as their predecessors! District transport offices were opened in Chelmsford and elsewhere in the region. The Road Service Licence was suspended and replaced by a 'Permit in lieu of RSL', also sometimes referred to as a 'defence permit'. The PSV Licence shown in Pic. 62 now incorporated the words 'Defence Permit'.

Clearly the war between Britain and Germany was going to affect both people and businesses, and Moore Bros. was no exception. First of all, due to the fear of Germany bombing cities — hence civilians — there was an urgent need to evacuate children from the capital. Whilst the trains would carry the heaviest burden taking the children to rural areas, buses would be needed to take them from the local railway station to outlying towns and villages where reception committees would be waiting in schools and community halls. Moore's, amongst many other operators in the area, were called upon to provide buses to fulfil this task. Some of Moore's buses were carrying out these duties in Colchester and as far afield as Hertfordshire.

An unexpected activity in which the company found itself was the camouflaging of the garage premises against detection from enemy aircraft. Not that the Luftwaffe regarded Moore Bros. as a strategic target, but rather that the recently erected canopy enclosing the courtyard in front of the bus sheds might be seen from the air and could attract unwanted attention. The canopy comprised corrugated asbestos panels supported on a steel framework. In the streaming sunlight of 1939 these new white panels could be quite distinctive from above and so a number of staff were employed in painting the panels by hand using brown and green paints. At the time of writing the camouflage paint is still visible, although now very faded.

We have seen how works services were introduced in the early thirties to convey the ever increasing numbers of people being employed in the local factories of Crittall, Marconi and Hoffmann. As these factories switched over to war production and the workforce expanded, there was a need to get these increasing numbers, now both men and women, transported to their places of work. This placed an even heavier demand on public transport and therefore created a need for more buses. By 1940 Moore's plight for vehicles had become desperate and, no doubt, given a choice of double-deckers they would have chosen AEC Regents, but presumably there were none available on the second-hand market. New manufacture of buses had almost ceased, for the industry was now engaged solely on war work. Later on, as we will see, shortages of passenger-carrying vehicles had generally become acute so the government were forced to recognise that bus manufacture for civil use was an essential commodity and plans were made to build them.

For the moment Moore's had to turn their attention to two second-hand Leyland Titans which had become available. A TD1 which had been in the service of Southdown Motor Services Ltd. of Brighton was purchased, as was a TD2 which had been with Hanson of Huddersfield. A picture of the Southdown Leyland whilst in service on Southdown's route 4 is shown in Pic. 46, whilst Pic. 47 shows the Hanson vehicle when new at the Leyland factory. This bus was fitted when new with one of the earliest 8.1 litre diesel engines, the type of which was still present when delivered to Moore's. Both these buses had 'highbridge' bodies — a centre gangway on the upper deck — which limited their use, but it does indicate the desperate need Moore's had for more vehicles. Much care would have been needed by drivers to take them through the railway arch at Kelvedon, with the one in Duke Street, Chelmsford, certainly being 'out of bounds'! These two double-deckers turned out to be very unpopular

46 Southdown Titan TD1. New 1930 to Moore's 1940.

47 Hanson Titan TD2. New 1933 to Moore's 1940 and first double-decker with oil engine.

amongst crew members, with heavy steering being one of the many problems, and no more Leylands were purchased. As noted in Chapter 8, by good fortune an AEC Regent had become available in 1941 which had been with South Wales Transport Co. Ltd., bringing the total number of Regents owned to four. The Moore directors must have been well satisfied with these purchases since we find an entry in the July 1941 issue of the 'AEC Gazette' that the company had advance ordered two Regent chassis for delivery when circumstances permitted. These latter words obviously referred to a time when AEC would be permitted to revert to civil manufacture. However, the course of events for Moore Bros. during the war forced a change in fleet strategy and no further AEC chassis purchases were made.

Soon after the start of hostilities in 1939 PSV chassis manufacture ceased on government instructions, with factory output concentrating on the war effort. Sudden stoppages such as this, like any branch of modern industry, would cause disruption to the flow of 'work in progress' causing a surplus of components which had been waiting chassis assembly. Of necessity, these would have to be put into store and 'frozen' until such time as they could be used. As demands from bus operators grew for new vehicles the Ministry of Supply were forced to recognise this need and therefore agreed to the material being 'unfrozen'. From this material a limited quantity of chassis could now be built. The allocation of these chassis to operators was under the strict control of the Ministry of War Transport to whom all applications for new vehicles had to be made. Very good reasons for acquiring one of the vehicles would have to be given.

It was from this source that Moore's received their only Bristol type K5G double-decker fitted with a 5-cylinder Gardner 5LW diesel engine. A picture of this bus appears in Pic. 48. Thus the chassis and engine were built to the manufacturers standard pre-war specification, but the fitted Duple body was to a strict 'utility' specification and a far cry from this manufacturer's normal standard of body construction. The utility body was designed by the 'National Federation of Vehicle Trades and Operators Joint Technical and Advisory Committee' under instruction from the Ministry of Supply and the Ministry of War Transport. The purpose of all this was to minimise use of materials, cut out all unnecessary frills and to de-skill construction. From an external point of view the body was angular in appearance with all rounded panels being dispensed with wherever possible resulting in a very austere look. A number of bodybuilders were employed in producing these bodies to the utility specification and therefore, not surprisingly, appearances were very

48 Bristol K5G with Duple utility body. Standing Top Road, Tolleshunt Knights.

48a New 1945 Guy Arab II. Moore's first double-decker with Strachans body. Duke Street, Chelmsford. 1950.

similar, although on close inspection subtle differences could be spotted. The interiors were very sparse looking, without lining to roof and side panelling. At first, upholstered seats were fitted but later on only wooden slatted seating was permitted. The rear upper deck emergency exit was now steel-panelled instead of being glazed, making the vehicle look very odd when viewed from the rear. Exterior painting, as delivered, was a matt grey finish, often referred to as 'battleship grey'.

The principle of adopting a standard specification for bodywork was applied later to chassis construction once the unfrozen material had been used up. It was in this way that Moore's were introduced to the products of Guy Motors of Wolverhampton — Guy Motors being one of only two manufacturers given authorisation to build bus chassis. An introductory page from a wartime Guy Arab service manual shows Guy Motors concern about damage to their reputation through manufacture of an enforced inferior product, see Fig. 20. Clearly they had no wish to lose the feathers in their cap and so destroy their motto enshrined for so long in the Red Indian mascot mounted on the radiator cap. The first Guy double-decker arrived in 1943 which was a Guy Arab I having a 5-cylinder Gardner 5LW engine — the power unit upon which Moore's were to standardise for most of their fleet until the sell-out to Eastern National in December 1962. The first Guy Arab was fitted with a body by Brush of Loughborough which was the only Brush-bodied example to come into the Moore's fleet, apart from one inherited on a second-hand AEC Regent. Of the remaining six Guy double-deckers to be delivered up to and just after the end of the war in 1945, one was bodied by Chas. H. Roe of Leeds, with Strachans of Acton, London, supplying the rest. These bodies were all fitted to the later Guy Arab II; the only difference from the Guy Arab I being a slightly longer bonnet to accommodate the 6-cylinder Gardener 6LW diesel engine which when ordered, was subject to allocation approval by the Ministry of War Transport and generally only granted when the route terrain demanded additional power. The longer bonnet necessitated a slight change to the front wings by adding an up-turn giving the appearance of a lip to the front edge.

For single-deck manufacture a similar approach was adopted by use of a utility body specification. Although some bodies were built for use on unfrozen chassis by various well known manufacturers, by far the largest body production was for the Bedford OWB chassis; the main body supplier being Duple of Hendon. Two only of this OWB chassis were ordered by Moore's which were bodied by Mulliner of Northampton. These were delivered in 1945, towards the end of the war in Europe.

Fig. 20 From 1942 Guy Arab Service Manual.

INTRODUCTION

GUY "ARAB" DOUBLE DECK OIL ENGINE CHASSIS

(War-time Production).

THE vehicle is built under the instructions of the Ministry of Supply to a specification agreed between them, the Ministry of War Transport and the Technical Committee of Operators.

The general design is based largely on the Guy "ARAB" pre-war model, which proved, upon investigation, to have given a very good account of itself, vehicles having run anything from 250,000 to 400,000 miles.

In the years that have elapsed since these vehicles were produced, naturally some detailed improvements have been made, but owing to war-time conditions the Company was instructed to eliminate all "frills." The use of elektron and aluminium had to be eliminated and this, together with other details, resulted in a considerable increase in weight on the chassis, an increase of approximately 18·5%, and on the complete vehicle (depending on the make of body), anything from 13·9% to 19·5%.

For the more difficult routes in view of this increase in weight, we should have preferred to have fitted a six-cylinder Gardner engine, but the M.O.S. advised us that the production of this type is very limited, and the majority of vehicles must be produced with the five-cylinder engine. This increase in weight, together with the much larger number of standing passengers allowed during the war, affecting as it does the power weight ratio, must inevitably detract from the performance of the vehicle compared with the pre-war models, and users are asked to bear this very important point in mind.

In the cases where vehicles get into the hands of drivers who have previously only driven buses with an epicyclic gearbox, may we point out that it is estimated that only about 1% of all the motor vehicles on the road have such a type of gear fitted, the other 99% are of the crash type box, which is successfully driven by the general public, including a large percentage of women drivers. In short, if persons cannot satisfactorily handle a crash type gearbox they cannot consider themselves drivers, a fact which we are sure the managements fully endorse. The simpler construction of the crash type box (which incidentally has fifty-nine fewer parts compared with the epicyclic) is of great importance in war-time from the point of view of manufacture and maintenance.

The particular gear ratios in the gearbox are rather different to those in the pre-war vehicles, and make for much easier changing, and this, in combination with the multiple clutch we employ (which gives a delightful take up) can in the hands of an average driver give the results as soon as any fluid flywheel and epicyclic gearbox.

Incidentally some of the pre-war Guys have not had a replacement in the gearbox or any renewal of the clutch fabrics after 400,000 miles.

The Company will, at all times, be happy to receive any suggestions with a view to improving the "breed" if war-time conditions allow, and we have every confidence that these vehicles will give long and satisfactory service, and any advice and assistance we can give to attain this end will be a pleasure.

The Bedford OWB type was the wartime version of the manufacturers OB type which had been introduced in 1939. Production of the OB resumed in 1946 and continued into late 1950 when it was replaced by the SB type chassis. As far as outward appearance was concerned the Bedford OWB radiator grille and bonnet was very similar to the WTB Bedford's delivered to Moore's in 1938, but the body, as with the double-deckers, was very angular in construction. Seating was of the wooden slatted variety and provided for 32 passengers. A wartime advert of the Bedford OWB is shown in Fig. 21.

From about 1942 utility bodies were made available for existing chassis and through this outlet Moore's decided they would re-body the three Regent double-deckers which had previously been with the City of Oxford. Presumably the original Hall Lewis bodies which had been built in 1930 were now in a very poor condition due mainly to plywood being used for panelling, and it was time to replace them. Although these buses would take on an austere look, the new bodies by Willowbrook of Loughborough and Northern Coachbuilders of Newcastle upon Tyne, gave them a new lease of life and they remained in the fleet until 1955. AEC Regent WL9068 with Northern Coachbuilders bodywork is shown in Pic. 49.

Fig. 21 Bedford OWB wartime advert.

49 AEC Regent with utility Northern Coachbuilders body in Roslyn Garage.

So far we have mentioned the difficulties at the commencement of war and problems with the expansion of the bus fleet. As the war deepened into greater conflict many other problems had to be faced and overcome by the company. Not least of these was the loss of staff on active war service. This was solved by the employment of conductresses, some of whom remained in Moore's service for many years after the war ended. The 'black-out' restrictions which came into force on 1st September 1939 created particular problems for drivers and conductors. Headlamps were fitted with masks which cut out spurious light and left only a small downward beam onto the kerb and a small patch on the road. As an aid to being seen at night all vehicles had to have a white strip painted around the outside edge. All interior lighting was masked which gave just a glimmer of light totally inadequate for the conductor to see to collect fares and issue tickets. To overcome the problem the conductor would have a torch attached to the ticket punch or money-bag strap. The Eastern National had the ingenious solution of a shaded lamp which could be slid along a copper strip fixed to the ceiling of the lower and upper decks of the bus.

Fuel rationing was introduced in the early part of the war for the private motorist and later this ration was cut off altogether, with cars having to be laid up for the duration of the war. This was to have the effect of increasing passenger traffic for the bus industry, who themselves were not immune from fuel restrictions as the crisis intensified. The first noticeable effect on the public was the discontinuance of day excursions and other pleasurable trips in September 1939, also the complete withdrawal of services on Christmas Day 1939. It is worth noting that from the early 1930's Moore's had provided a service on Christmas Day up to lunch time, but from what started as an enforced cancellation this service was never reinstated. By 1942 the crisis had further deepened so that there were no services before lunch time on Sundays and the last through bus from Chelmsford to Colchester during the week was 6.45 p.m. with a later departure of 7.45 p.m. on Saturdays and Sundays. Similar departure times applied to through services from other towns. No departures were permitted after 9.00 p.m. This was a far cry from the late departure times which had become a feature of the 1930's, principally for theatre and cinema goers.

Later on further measures were introduced like giving workers and schoolchildren, who were holding weekly, monthly or quarterly tickets, priority travel over other passengers. Restriction on the permitted number of standing passengers was eased to increase the carrying capacity of vehicles. Also, compulsory queuing had to be introduced to bring some order to waiting passengers. For many people, queuing became a way of life and it was often said if a queue had formed it was best to join it even if the purpose was not known — sometimes there could be a surprising result!

During the early part of the war travelling to Eastern Coastal areas was very restricted due to the presence of heavy fortifications and troops, to guard against the threatened invasion by German armed forces. Checkpoints manned by British troops were a familiar sight on many roads where an identity card had to be produced when requested. Absence of an identity card could result in detention. One such incident involved Hubert Newport who having journeyed to London with a party from Maldon in Moore's Albion BEV465, left the city, with the party's agreement, during a heavy bombing raid. Having safely delivered the party to Maldon he then headed for Kelvedon via Great Braxted where an army troop was encamped in the park. The bus was stopped at the checkpoint and as the driver didn't have his identity card with him the bus was seized and Hubert Newport was locked up in the guardhouse for the night, just $1^1/_2$ miles from home. He was released the following morning once Harold Moore was able to vouch for the bus and driver!

At the height of the Battle of Britain it was not uncommon for passengers to be asked by the driver and conductor to leave the bus and take cover in the nearest ditch when a dogfight was taking place overhead. During 1944 and when preparations were being intensified for the Allied invasion of Europe, the general public were restricted to travel within five miles of their home unless they possessed a permit to travel to a specific place or places. This could be a hindrance to the bus crew when a police officer would board a bus to check the passengers' identity cards or permits. Anyone not sanctioned to continue the journey would be ordered off the bus with instructions to return by the next bus from whence they came!! Of course, all of this took time much to the annoyance of passengers, especially those unable to complete their journey, and also to the crew members. To everyone's great relief 1945 saw the end of hostilities in both Europe and the Far East, but it would be some considerable time before rationing and many other restrictions would end which could allow bus services to resume their normal peacetime role.

Chapter 10

Post-war Adventures

With the end of World War II there was much rejoicing with Victory parades, street parties, and many other celebrations taking place throughout the country. In the immediate post-war period shortages of materials continued to impose restrictions on people's social lives, none more so than that of petrol rationing, and with all new car output allocated to those with an essential need the growth of private car ownership was still a long way off. This caused second-hand car prices to soar, including those vehicles which had seen better days and under different circumstances would have been on their way to the breaker's yard.

All of this should have been good news for the bus companies but they in turn had their own problems. One good illustration of this being where Moore's were unable to resume their pre-war programme of excursions and tours from Kelvedon and Coggeshall until as late as the summer of 1950. Availability of buses was still very restricted and those being built were to a utility specification, although the end of 1944 brought a relaxation to the body specification. From a passenger viewpoint the most important items were a return to upholstered seating, and two opening windows being allowed instead of one on each side of the lower and upper decks. Outwardly there was little change to the body apart from the roof line where this had been softened by the use of rounded roof domes instead of the earlier angular form. Moore's were to receive a number of double-deckers with this relaxed specification, all built by Strachans of Acton, London. This was a return to a body builder who had supplied single-deck bus bodies on Thornycroft chassis in 1929, although at that time the firm was known as Strachan and Brown. However, this earlier association had been very short-lived, and was before Moore's had decided to use Duple bodywork for all their new coaches and buses in the 1930's.

Strachans was now to be the supplier of Moore's first double-deck bodies of post-war design, delivered in 1947. These bodies were of an improved design and quite a refreshing change from the drabness of the wartime versions, and especially those earlier second-hand purchases of AEC and Leyland with outdated bodywork styles. The appearance of these two double-deckers, LPU611 and LPU612, was further enhanced by the fitting of these bodies to the Guy Arab III chassis — which first appeared in late 1946 — where the radiator and bonnet line had been lowered by as much as four inches. This version of the chassis was available in both single- and double-decker form and with the lifting of government restrictions on the use of alloys in chassis manufacture, a reduction in overall weight was possible. Apart from the 5 and 6-cylinder Gardner diesel engines being available, the more powerful 6-cylinder Meadows diesel engine could be fitted as an option. In fact this latter engine was fitted to four Guy saloons which Moore's purchased second-hand in 1953 from Dartmouth Coaches, who were based in London. These engines would have been ideal for excursion work but in the interest of standardisation and economy they were later replaced by Moore's with their standard 5-cylinder Gardner 5LW unit, two of which came out of the ex-Hicks chassis, as we shall see in the next chapter.

The Guy Arab III chassis was to figure prominently in the Moore's fleet over the next few years with Strachans bodywork being used for most of the single-deckers and all of the double-deckers. The arrival of the first two Guy Arab III double-deckers signalled the end of the South Wales AEC Regent which had not been re-bodied during the war, unlike the other three AEC Regents. The infamous Leyland Titans had been disposed of in 1946.

With the introduction of peacetime designs in bodywork it was now time to consider replacement for the Albions which had given such faithful service since their first arrival in 1934. The intervention of World War II and the resulting restrictions on vehicle manufacture introduced Moore's to Guy Motors Ltd., as a supplier of double-decker chassis. They were now to use this manufacturer for their single-decker chassis, but initially they remained loyal to Duple for bodywork. The first four saloons, LPU940, LTW62, LVW124 and LVW125, were dual-purpose having 35 seats and were delivered in 1947. The interiors had lost the luxury feel of their 1930's counterparts, and whilst their purchase had been intended for both stage carriage and excursion work, the lack of power of the 5-cylinder Gardner engines really made them unsuitable for long distance work in hilly terrain. The first Albion disposals were the two Victors, DNO837 and EVW868, which passed to Tye Bros. of Mendelsham, Suffolk, who were later taken over by A. F. Braybrooke and Son. A picture of DNO837 whilst in the livery of Braybrooke's can be seen in Pic. 50. The remaining Victor — also the petrol-engined Valkyries — were sold during 1947.

50 Ex-Moore's Albion Victor with Braybrook, Suffolk.

50a 1947 first Guy Arab III Duple saloon.

Fig 22 1948 timetable.

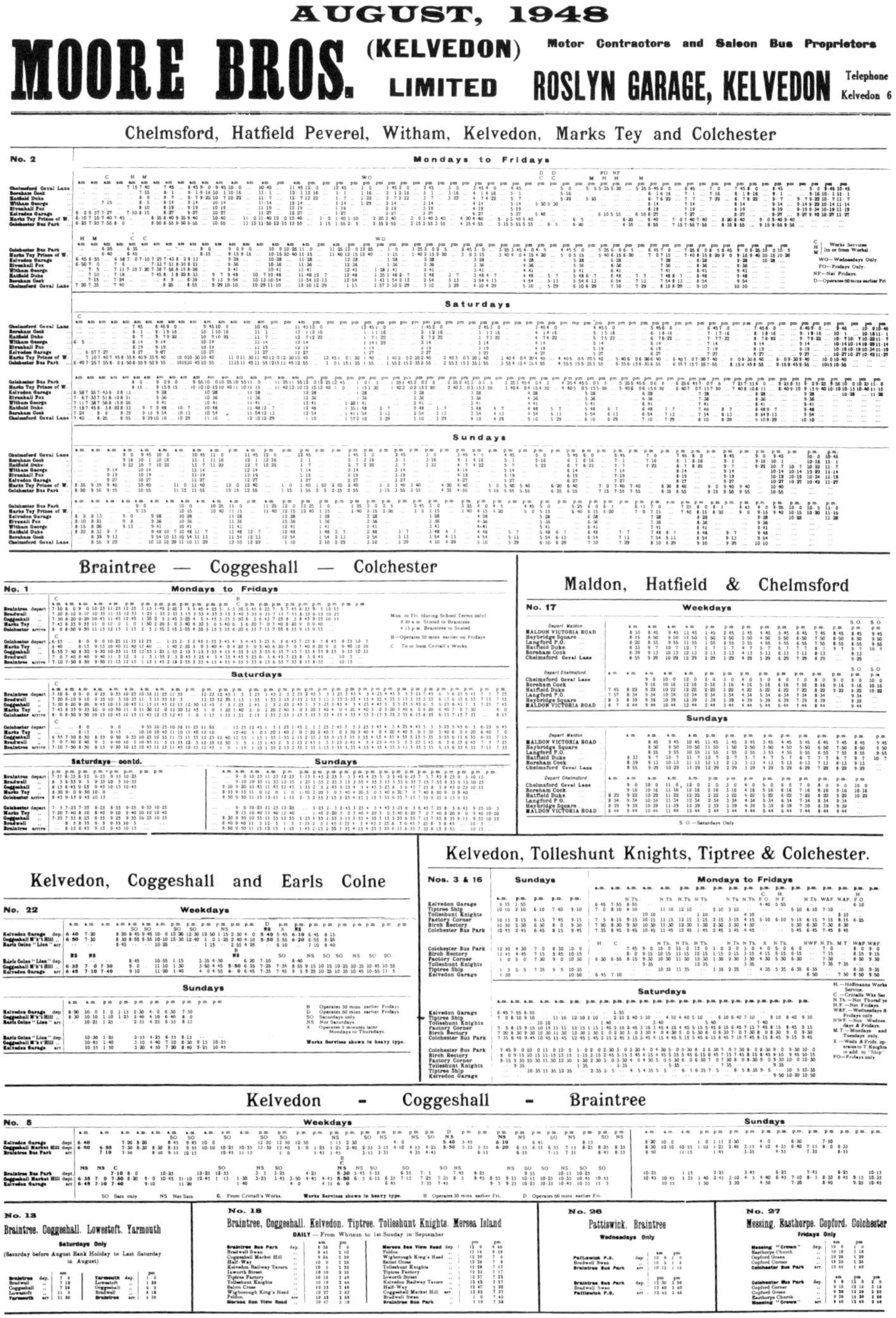

AUGUST, 1948

MOORE BROS. (KELVEDON) LIMITED

Motor Contractors and Saloon Bus Proprietors

ROSLYN GARAGE, KELVEDON

Telephone Kelvedon 6

Chelmsford, Hatfield Peverel, Witham, Kelvedon, Marks Tey and Colchester

No. 2

Mondays to Fridays

Saturdays

Sundays

Braintree — Coggeshall — Colchester

No. 1

Mondays to Fridays

Saturdays

Saturdays—contd.

Sundays

Maldon, Hatfield & Chelmsford

No. 17

Weekdays

Sundays

S.O.—Saturdays Only

Kelvedon, Coggeshall and Earls Colne

No. 22

Weekdays

Sundays

Works Services shown in heavy type.

Kelvedon, Tolleshunt Knights, Tiptree & Colchester.

Nos. 3 & 16

Sundays

Mondays to Fridays

Saturdays

Kelvedon - Coggeshall - Braintree

No. 5

Weekdays

Sundays

SO Sats only NS Not Sats. C From Crittall's Works. Works Services shown in heavy type. B Operates 30 mins. earlier Fri. D Operates 60 mins. earlier Fri.

No. 13

Braintree, Coggeshall, Lowestoft, Yarmouth

Saturdays Only

(Saturday before August Bank Holiday to Last Saturday in August)

No. 18

Braintree, Coggeshall, Kelvedon, Tiptree, Tolleshunt Knights, Mersea Island

DAILY — From Whitsun to 1st Sunday in September

No. 26

Pattiswick, Braintree

Wednesdays Only

No. 27

Messing, Easthorpe, Copford, Colchester

Fridays Only

While every endeavour is made to operate this Time Table according to schedule, the operators cannot hold themselves responsible for connections missed through unforeseen circumstances

But we must now return to 1946, a year that saw a significant development in the service network. From the earliest days of the carriers cart, and the start of the horse-drawn omnibus in 1881 on the Kelvedon to Coggeshall route, the Moore family had regularly faced competition on most of their services, but now suddenly Hicks Bros. of Braintree decided to withdraw from the Braintree to Colchester service by surrendering their Road Service Licence, leaving the road entirely to Moore's. Why Hicks Bros. had decided to do this is unclear, but it is quite evident that they were ready to reach an agreement with Moore Bros. that in exchange for withdrawing from this road Moore's were to give up using Braintree as a centre for their excursions and tours. Any such agreement would of course be subject to the Traffic Commissioners approval on the transfer of the licences. This was agreed and the details of application and grant appeared in Notices and Proceedings, Nos. 416 and 420, of September and November 1946. The beginning of 1947 also saw licences being granted to Moore's to operate two new stage services. The first was between Messing (The Crown) and Colchester which was to run on Fridays only and offered three return journeys via Easthorpe, Copford Green, Copford Corner and thence via the main A12 through Lexden to Colchester. Much later on this service was combined with that of the Kelvedon to Earls Colne service giving a through route from Colchester to Earls Colne. The second was between Pattiswick (PO), Bradwell (The Swan), along the A120 to Braintree, which was to operate on Wednesdays only and provided two return journeys. These services started on 7th and 5th March 1947 respectively. To some extent these new services could be viewed as Moore's replacements for the two services; Kelvedon to Braintree via Earls Colne and Halstead, also the Earls Colne to Maldon service, both of which were withdrawn at the outbreak of war in 1939.

The year 1949 saw the arrival of two new very smart Strachans-bodied double-deckers, ONO537 and ONO538. These were Mk III Arabs and the last new double-deckers to be delivered to Moore's with exposed radiators. It is interesting that two almost identical vehicles were delivered later in the same year to Hicks Bros., OVW756 and OVW757, which were to end up on Moore's premises following disposal by the Eastern National in 1959. We shall return to these second-hand buses in the next chapter. The delivery in 1949 of the two attractive looking double-deckers must have convinced the Moore directors that their decision to use Strachans bodywork for their new Guy Arab III saloons had been the right one. Thus four Strachans 35-seater single-deck saloons, PHK531 to PHK534, arrived during late 1949 and early 1950. However this was not the end of Moore's association with Duple since their coachwork was to be fitted to a new 27-seater Bedford Model OB saloon, RTW222, for delivery in January 1951. Normally these saloons were fitted with 29 seats, but 27 seats permitted them to be used on stage work without a conductor. This increased versatility was to prove very useful on sparsely populated routes, as well as being available for relief work and contract hire. Bedford OB bus/coach production ceased in November 1950 and so this was not only one of the last to be built, but it benefited from a more powerful 6-cylinder petrol engine which had been introduced in March of that year.

All of these new acquisitions not only modernised the fleet but also allowed progressive disposal of the pre-war Albions and Bedfords. Whilst this was highly desirable in the interests of standardisation and reduction in major overhauls, a gap still remained in carrying capacity during the early 1950's and so, once again, attention was turned to the second-hand market. Clearly the new purchases were putting a strain on the vehicle budget and a balance had to be struck between the need to increase fleet size and the expenditure needed to achieve it. Also, buying second-hand had the advantage that provided the right vehicles could be found from dealers, the long lead periods which manufacturers required for new buses could be avoided.

50b First really smart Strachans double-decker. New 1949. Braintree Bus Park 1950.

50c The taste for Strachans bodies now spread to saloons. St. John's Bus Park, Colchester. c1950.

The first second-hand acquisitions to arrive in January 1951 were two 29-seater Bedford Model OB's with Duple Vista bodies which had been in the fleet of Fallowfield and Britten, who were coach operators in North East London. One of these, MMT876, had a shortened life with Moore's as it became involved in a serious accident on 28th August 1954 and was subsequently scrapped. A newspaper report and picture of the accident can be seen in Pic. 52a. The other Bedford, HYM685, is shown in Pic. 51. These were followed during the latter part of 1951 and early 1952 by four Guy Arab II double-deckers, GYL981 to GYL984, which had been in the service of Birch Bros. Ltd. These carried Weymann utility bodywork and the buses were new to Birch Bros. in 1945. A picture of GYL984 at Henlow Camp, whilst in the service of Birch Bros. Ltd., can be seen in Pic. 52. All four Guys were purchased to operate the Henlow Camp to Luton service. An indication of restricted financial resources at this time can be gauged by the decision to buy these buses second-hand and yet within two or three years of this happening a re-bodying programme had been embarked upon, confirming that short-term expediency had been crucial.

51 Bedford OB on Kelvedon - Earls Colne service.

52 At Henlow Camp, Beds.

52a Serious accident at Woodham Mortimer. MMT 876.

The following was reported in the Colchester Gazette of August 31 1954

20 passengers hurt in coach crash

Twenty passengers from the Kelvedon area were hurt when a coach returning from the Rayleigh speedway meeting was involved in a colision with an Army recovery vehicle at Woodham Mortimer on Saturday night.

The coach had half of one side and the roof ripped away and the Army vehicle caught fire and was damaged.

Eleven of the more seriously injured coach passengers were taken to the Essex and Chelmsford and St. John's Hospitals, Chelmsford, for treatment, after emergency treatment at Maldon.

EYE OPERATION

Mrs. Joyce Challis (21), of Popsold Road, Kelvedon, had an eye operation at the Essex and Chelmsford Hospital. She sustained a lacerated scalp and face.

Mrs. Dorothy Townsend (24) of Coggeshall Road, Feering, was also detained with head injuries.

The coach driver, Mr. Charles Smith, of Kelvedon, was badly shaken, but unhurt.

A passenger in the Army lorry, Cpl. Frederick Pearce, of 6 Camp, R.A.S.C., Orsett, had rib injuries, and Mr. Russell Sparrow (26) of St. Mary's Road, Kelvedon, who had lacerated face and scalp, were both detained at St. John's Hospital, Chelmsford.

Firemen from Maldon and Cold Norton fought the lorry blaze.

52b GYL983 St. John's Bus Park, Colchester. c1953.

53 Rebodied by Strachans. Coval Lane Chelmsford. c1955.

It was not uncommon to find some wartime and even post-war bodies suffering from rot due to the use of unseasoned timber for the framework. This called for repair work, which sometimes became very uneconomic and as body manufacturing capacity increased re-bodying existing chassis became a possibility. For Moore's garage staff the re-bodying of the four ex-Birch chassis was quite an extensive task and involved the removal of the existing body, overhauling the chassis and fitting a replacement engine before driving each one of them to Strachans of Acton, for a new body to be fitted. Whilst the new body was a vast improvement on the utility version, the vehicle did take on a slightly strange appearance due to the 8 foot wide body being mounted on a 7 foot 6 inch chassis, which resulted in the wheels being well inside the wheel arches. A picture of GYL983 with the new Strachans body is shown in Pic. 53 whilst on layover in Coval Lane, Chelmsford. Although the re-bodying programme was to be a proven success, from 1954 onwards all double-decker additions to the fleet would be new, apart from an interesting re-bodying episode which took place during 1960.

54 Approaching Durwards Hall on A12, near Rivenhall.

54a Northern Counties bodies were fitted to only two of Moore's Arab IV chassis. Duke Street, Chelmsford.

Excursions and private hire work had always formed an important part of Moore's business, especially since the arrival of the first motorised bus, which you will recall was a second-hand 14-seater Daimler purchased specifically for that purpose. As if to continue the tradition, four Guy Arab III saloons with Strachans coachwork, KGT378 to KGT381, were purchased second-hand in 1953, having previously been with Dartmouth Coaches in South-East London. With the progressive ending of wartime rationing and a general feeling that life was improving, the desire to travel further afield was placing increasing demands on coach operators. School outings and holidays were also becoming popular. Those fortunate enough to travel in these coaches in the early years were treated to an experience never before known on Moore's buses! Radio, heating, forced ventilation, and a powerful 6-cylinder Meadows engine which was well able to keep up with other express coaches of the day — the embarrassment of long lines of coaches overtaking, especially on hills, was a thing of the past, at least for a time! As already mentioned, in the interests of economy and standardisation the Meadows engines were removed over a period of time and replaced by the 5-cylinder Gardner 5LW units.

However, all these new developments were now overshadowed by the death of Horace Moore on 15th January 1955 at 83 years of age . Horace had become Managing Director following the death of his brother, William, in 1940. I am sure most people will remember him for always maintaining an air of authority, and for his correctness — also an expectation of this from others — and for being immaculately dressed. I can easily recall the incident when Eastern National were running late on the Chelmsford to Colchester service so that they arrived in Kelvedon just a few minutes ahead of the Moore's scheduled arrival time. Horace Moore standing outside his garage as always, timing his buses in each direction, did nothing more than stop the Eastern National bus and question the driver as to why he was running to Moore's timetable. Whatever answer was given Horace demanded he wait until the Moore's bus had passed through at its scheduled time, which it duly did. Then the Eastern National bus was 'released' to continue its journey. In spite of the rivalry between the two companies a level of respect remained.

Earlier on we said the Arab III double-decker was the last of Guy Motors standard production to have an exposed radiator. From now on radiators would be concealed by an ornamental front, or 'tin-front' as they became commonly known. The design, embodying a new style of bonnet and front wing, changed the frontal appearance altogether. The idea had originated as part of a Guy Motors development to meet the requirements of Birmingham City Transport. With some small modification it later formed part of standard Guy chassis manufacture. This became known as the Arab IV. The first arrival at Moore's with this design was Strachans-bodied double-decker, YEV263, in 1954, see Pic. 54.

Since the end of World War II in 1945 Moore's had used Strachans exclusively to body their double-deckers, but now in the mid-fifties this body builder had decided to withdraw from the double-decker market. The last Strachans-bodied vehicles to be received were 52DHK and 53DHK, in 1955. The company had now to look elsewhere for a body builder and turned their attention to Northern Counties and Engineering Company Ltd. whose works were situated in Wigan. By now Guy Motors had introduced their 30-foot long double-decker chassis which gave

55 Moore's first Commers and full fronted coaches shortly after arrival from Duple.

56 Evening rush hour. Victoria Road, Chelmsford. Eastern National strike 1956.

greater seating capacity, but at the expense of increased overall weight. Some small reduction in weight was achieved by using fibreglass instead of metal for the bonnet top, front wings and concealed radiator panel, but increased power was desirable for this larger bus and therefore Guy Motors decided they would not offer the 5-cylinder Gardner 5LW unit for this chassis version. However, it is interesting to note that an exception was made for Moore Bros. and the 5-cylinder engine was fitted, principally because of their services operating over a flat terrain. These two buses, 2834F and 2835F, were delivered in 1957. Moreover, through this decision of Guy Motors, the 5-cylinder Gardner unit would continue to be made available to Moore's on these larger buses. After using Northern Counties to body only two of their Arab IV double-deckers, all future orders were to be placed with Massey Bros., also of Wigan. By a strange twist of events, in 1967 Northern Counties took over Massey Bros. Ltd., whose works were at Pemberton, the premises of which Northern Counties would continue to use. All eight Massey-bodied double-deckers ordered by Moore's were fitted onto the 30-foot Arab IV chassis with a 67 seating capacity.

When Bedford OB production ceased in November 1950 it was replaced by the Model SB, most of which carried Duple Vega coachwork. Although the Model SB chassis never entered Moore's fleet, the Vega body was chosen for fitting to two Commer Avenger Mark III chassis delivered in 1956. Thus we have a return to Duple — but for the last time. These coaches, 576FVW and 577FVW, were obtained exclusively for excursion and private hire work. Many who travelled on these coaches may well remember the 'bark' that emitted from the two-stroke diesel engine — quite a contrast from the 'singing' gearboxes of the Bedfords. These were the first full-fronted coaches to enter Moore's fleet, see Pic. 55.

This chapter has shown perhaps more than anything else the way in which Moore's were able to expand their vehicle fleet to cope with ever increasing numbers of passengers. This was no mean feat bearing in mind the company was still a close-knit family business, with just four shareholders, and limited finances.

The stage carriage traffic was at its most intense at the start and finish of the working day. It was quite common to find eight double-deckers passing through Kelvedon between 6.30 and 8.15 am en route to Witham and Chelmsford, some of which were directly serving the factories of Crittall, Hoffmann and Marconi. Certain of these buses would then return to Hatfield Peverel to serve as 'reliefs' for the Colchester and Maldon services to Chelmsford. Moore's also held a number of contracts for schools that placed a demand on buses as soon as the peak morning period was over. Obviously the reverse took place in the late afternoon and evening. An indication of the number of buses leaving Chelmsford during the peak evening rush can be seen in Pic. 56 This was taken during the Eastern National strike of 1956 when Moore's were advised not to proceed further into Chelmsford than Victoria Road. Consequently all their bus departures were from Victoria Road instead of Coval Lane.

The summer service from Braintree to West Mersea was very popular during the 1950's when, on a Sunday the garage could be found deserted, with just the little Bedford 'breakdown truck', CHK401, in evidence. The extra buses were pressed into service at Kelvedon because the regular vehicles were already full on arrival. The number of buses required could never be pre-determined, of course, because being a stage carriage service there was no advance booking. If it was a nice sunny day people just turned up, with the queue getting longer and longer. Who could deny such a happy crowd from enjoying a few hours at the seaside, especially the children with their buckets and spades? Finding space for these people was one thing, but what about those who would be waiting at Feering, Inworth, Tiptree and Tolleshunt Knights? It had been a long held tradition in the family business that passengers did not get left behind. It was this that drove the staff to find the extra seating capacity needed. To crew the buses the garage staff were brought into service, and even an inspector, if one was available.

The Summer Saturday express service to Lowestoft and Great Yarmouth from Braintree continued to flourish during the whole of this period — to a point where the service was extended to The Silver End Hotel in 1952. An attempt was made at this time to include a further picking-up point at The Prince of Wales, Marks Tey, but this failed. Shortly after being granted the licence to operate this service back in March 1932, Moore's applied to run express services to Clacton-on-Sea from Chelmsford and Braintree, but these were refused. It is interesting to speculate on how successful those services might have been had the licences been granted.

Chapter 11

The End Of An Era

56a First Guy Arab III Strachans body new 1947. St. John's Bus Park, Colchester.

As we enter the 1960's the Moore directors could well look back on their business performance over the past decade with great satisfaction. During these years significant numbers of new vehicles had joined the fleet and because of an ever increasing demand on carrying capacity some earlier vehicles had been modernised and were no longer looking out of place. Both these policies would continue to be applied during the 1960's. To an outsider it would seem that with purchases of new double-deckers becoming an almost yearly event, the company could now embark on a disposal programme for the older vehicles in the fleet. However, this was not to be the case, apart from the Bristol double-decker, JPU640, whose disposal was not surprising since it had become the sole remaining make of chassis in an otherwise Guy Arab double-decker fleet. In fact, far from a disposal of buses the reverse was to take place when three of the earliest Guy double-deckers received a chassis overhaul and had new bodies fitted. This work was to be undertaken during 1960.

56b Coval Lane Chelmsford. LPU611 rebodied Massey 1960.

57 Moore's first Guy double-decker with Brush bodywork new 1943. Note front opening windows now sealed to prevent leaks and new 'fresh air' window at side. Duke Street, Chelmsford 1950.

57a Two Hicks Guy Arab III purchased by Moore's 1959 for spares. Chassis OVW757 used in rebuild JTW447 in 1960.

58 Railway arch Duke Street Chelmsford c1961. Massey body 1960. Note lower radiator line.

As we have seen in Chapter 10, the exercise with the ex-Birch buses had been a success and showed that old vehicles could be made to look modern and perform reliably. The first to be taken in for rebodying was the Guy Arab III, LPU611, delivered to Moore's in 1947. Strangely its twin, LPU612, was not chosen as part of this exercise. Instead, two even earlier examples were taken — an Arab I, JTW447, delivered in 1943 and an Arab II, JVW999, delivered in 1944. In making this decision the problems that were to be encountered by the garage staff on these latter two chassis clearly had not been foreseen. On removal of the bodies it was discovered that the chassis frames were beyond economic recovery. However, by a strange twist of irony this became a blessing in disguise since it allowed the chassis from the two ex-Hicks buses to be used, which were Arab III having a lower radiator and bonnet line. These features would help to give an even more modern appearance following rebodying by Massey Bros. and reference to Pics. 57 and 58 show the contrasting styles to good effect. As mentioned in the previous chapter, the two ex-Hicks double-deckers had been purchased from Eastern National in 1959. The disposal of these two buses arose from their failure to gain Certificates of Fitness from the Ministry of Transport Examiner whilst in the Eastern National

59 Commer Avenger IV. Yeates body new 1960. Wembley Stadium.

fleet. However, as failure was due to body defects this was of no consequence to Moore Bros. who had only been interested in the engines and chassis for spares. To avoid any possible registration number problems the chassis identification plates were transferred from the now scrapped Moore's chassis to the ex-Hicks chassis. Rebodying then followed the same pattern as the ex-Birch exercise, with the Moore's garage staff having the task of removing the old body, overhauling the chassis and fitting a replacement engine, followed by driving each chassis in turn to Massey's premises at Pemberton in Lancashire for a new body to be fitted. The ex-Hicks Gardner 5LW engines had already been used to replace the Meadows units in the last two of the ex-Dartmouth coaches.

A surprise arrival during 1960 was a further Commer Avenger luxury coach, but this time with a Mk IV chassis and carrying a body by Yeates of Loughborough. The 'Europa' body style was very flamboyant and considered by many to be rather garish. As this was the last coach to be taken into the Moore's fleet it was unfortunate it attracted such adverse comment. The coach, 7652VW, was procured, it was said, primarily for taking the Colchester United football team to their away fixtures — Moore's having secured the contract for this work. The coach was also available for private hire, and can be seen on such a trip to Wembley Stadium in Pic. 59.

Although attention continued to be given to the excursion and tour side of the business no expansion to the stage carriage network had been seen since 1947. To some extent this was understandable since with the

Moore Bros. (Kelvedon) Ltd. Roslyn Garage, Kelvedon Telephone : Kelvedon 306

COMMENCING DECEMBER 5th, 1960

No. 32

TIPTREE . HAYNES GREEN . BLACK PITS . COLCHESTER

Mondays to Fridays

		a.m.
Tiptree Factory Corner	Dep.	8.15
Haynes Green Cross Roads	,,	8.20
"White Horse"	,,	8.23
Heckford Bridge	,,	8.32
Black Pits	,,	8.37
Villa Road (A12 Junction)	,,	8.40
Colchester Bus Park	Arr.	8.49

		p.m.	p.m.
Colchester Bus Park	Dep.	3.55	5.55
Villa Road (A12 Junction) ...	,,	4.04	6.04
Black Pits	,,	4.07	6.07
Heckford Bridge	,,	4.12	6.12
"White Horse"	,,	4.21	6.21
Haynes Green Cross Roads ...	,,	4.24	6.24
Tiptree Factory Corner ...	Arr.	4.29	6.29

No Service Bank Holidays

Fig. 23 Timetable covering last new service, no. 32.

constraints imposed by bus regulation there was very little opportunity for route expansion by the independent operator. Nevertheless, one such opportunity did arise which the company was able to fulfil towards the end of 1960. From the time of the steam bus, Moore's had run a regular service along the main Tiptree to Colchester road, and also a service to Colchester via the small villages that lay between this road and the A12. So, apart from the main road bus another bus could suddenly appear on this road only to vanish as it continued its journey along the back roads. As if to extend this tradition, an application was made to the Traffic Commissioners in 1960 to operate a service along a route that lay on the other side of the Tiptree main road. This would run from the Factory Corner via Haynes Green, The White Horse at Smyths Green, Black Pits, along Villa Road to the A12, thence Lexden to Colchester Bus Park. It was to operate Mondays to Fridays with one journey to Colchester in the morning, but two return journeys, one in the afternoon and one in the evening. The licence was granted and the service commenced on 5th December 1960.

With the company having standardised on chassis provided by Guy Motors apart from a few exceptions, they now realised they could be faced with a serious problem. Unfortunately during 1960 Guy Motors Ltd. went bankrupt and were now in the hands of the receiver. However, in October 1961 the business and assets were acquired by Jaguar Cars Ltd. and the entire Guy range continued to be produced and developed under the new

59a Fully loaded for Braintree. 20PVX leaves St. John's Bus Park Colchester.

59b 372WPU. One of a pair with 'Johannesburg' front.

name of Guy Motors (Europe) Ltd. Moore Bros. having had two double-deckers on order during the period must have breathed a sigh of relief at this outcome. The two Arab IV chassis with Massey bodies arrived in June 1961, and were fitted with a concealed front, but now with quite different styling. It became known as the 'Johannesburg' front; the name coming from a batch of buses supplied to Johannesburg Municipal Transport in South Africa having this style of radiator front. This particular style was regarded by many as the most appealing of the Guy designs, which by coincidence were fitted to the last two double-deckers delivered to the company, 372WPU and 373WPU. Two more Arab IV chassis were ordered in 1962, by which time Guy Motors had almost reverted to the original style of concealed front. Due to the dramatic events which were to take place in 1962 these vehicles, 581AOO and 582AOO, would not be delivered into Moore's ownership.

Whilst 1962 started as another year of great promise with two more new double-deckers being ordered on 1st April that year, the directors of the company, Harold and Percy Moore, who were also cousins, were now beginning to reach the twilight of their working lives. Before too long they knew they would be faced with a difficult decision. Both had inherited a business with a long family tradition which had been started by their great-great-grandfather, but now there were no descendants to carry it forward to the sixth generation. Percy's only son, Kenneth, had joined the Air Ministry soon after leaving school, making that his career, and displayed little interest in the bus company. Harold's only daughter, Patricia, had married a farmer and played no active part in running the business. There was also the efficient and loyal workforce to be considered. One or two employees still with the company had clocked up over 50 years of service, while others had devoted almost as many years. To add to this were the travelling public who had always been on very good terms with the company's bus crews. And so it can be seen that Harold and Percy had a flourishing company with a happy and contented workforce and an excellent public relations image, but they had no successors to the management.

Ever since the formation of The Eastern National Omnibus Company there had been countless rumours about that company wanting to buy out Moore Bros. and take them off the road. It can be assumed that these approaches would have been hotly rebuffed since the Moore family temperament preferred running their bus company well rather than exchanging it for cash. The family had a tradition of working hard for their living and expected their employees to follow suit, but not without reward in one form or another. However, during 1962 the Moore Bros. directors were again approached by Eastern National with an offer to purchase the business. This time the circumstances were different. The details of negotiations are not in the public domain, but ultimately agreement was reached and a draft contract drawn up between Moore Brothers (Kelvedon) Limited and The Eastern National Omnibus Company Limited. No doubt the document was modelled on past Tilling Group practice — Eastern National originally having been owned by Thomas Tilling. Solicitors acting for Moore Bros. were Bawtree and Sons of Witham — whose offices had been opposite Roslyn Garage in Kelvedon for many years, where they served as the firm's legal advisers.

Although the following is not intended to cover the full extent of the sale agreement, it does show some of the more important and interesting parts of it. The final agreement was signed on 31st December 1962 having a completion date set for ten days later, subject to the Traffic Commissioners granting the transfer of the necessary Road Service Licences. Some margin had been provided on the completion date, when the agreement would only become null and void if the licence transfer was not granted within three months of signature. The agreement covered the sale of:-

The freehold property forming the Omnibus Depot known as Roslyn Garage.
The goodwill business of Moore Bros. (Kelvedon) Ltd.
All the rolling stock comprising 43 vehicles, one of which was a chassis only.
Spares stock, plant, equipment and stores.

The total purchase price agreed for these items amounted to £85,000. This approximates to a purchasing power of £954,000 in 1996. The contract was also conditional upon the vendors not being part of, or employed by, any company or firm engaged in stage carriage, express, tours, excursions or private hire services — including the business of motor bus or charabanc proprietors — for a period of ten years, in the counties of Essex, Hertford, Suffolk, Norfolk or Kent. (no doubt a normally necessary restriction, but it scarcely applied to Percy and Harold Moore in the circumstances!). The contract for two new Guy double-deckers which Moore Bros. had entered into with the manufacturers on 1st April 1962, was assigned to Eastern National. These two buses were not due for delivery until after the takeover, which gave Eastern National an opportunity to request Massey Bros. to make bodywork changes. Thus these two new vehicles, 581AOO and 582AOO, were then not only finished in Eastern National livery, but the seating moquette was altered to conform to Eastern National practice and a service route box was added to the front destination panel.

During the later part of the sale negotiations rumours again started to circulate that Moore's were about to sell out to Eastern National, but as this had been heard many times before it was not taken too seriously at first. However, some staff were becoming increasingly concerned about this talk and after making enquiries in December were informed by the directors that negotiations were taking place. Not surprisingly word of such disclosures spread like wildfire, with villagers and passengers being stunned by the news. 'Taking to the streets' as a form of protest in the depths of winter was not an attractive option, even if such measures had been contemplated!! Even so feelings were running high, particularly amongst the staff and their families. A way of life appeared to be drawing to a close. These sentiments were to be resurrected some thirty years later as the next chapter will show.

During December 1962 Eastern National applied to the Traffic Commissioners for the transfer of all Moore's Road Service Licences. Because transfer had not been granted by the scheduled takeover date of 3rd February 1963, short period licences from this date until 30th March 1963 were issued. However, transfer was finally granted as confirmed in N&P's during April and May. Notice of the surrender of all Moore's licences was also given in the same N&P's. All the rolling stock which was transferred into Eastern National ownership is shown in Appendix VI. However, not all vehicles taken over were to be operated by Eastern National, but more of this in the next chapter. The last day of services running under Moore Bros. ownership was 2nd February 1963, with Eastern National taking over the company the following day. In common with Eastern National practice a free leaflet was issued for passenger information stating that Eastern National would be operating the services of Moore Bros. from 3rd February but there would be no changes to either time-or fare tables. Also, as Eastern National made use of route numbers on vehicles and in their timetables, these numbers were shown against each of the ex-Moore's services. The front page of this leaflet is shown in Fig. 24.

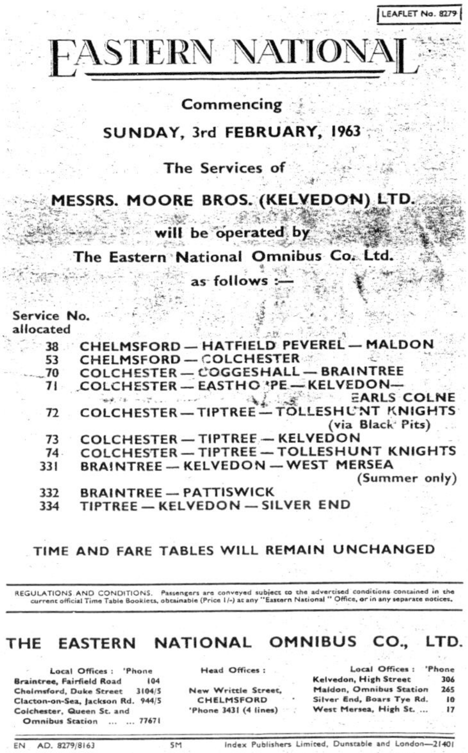
LEAFLET No. 8279

EASTERN NATIONAL

Commencing

SUNDAY, 3rd FEBRUARY, 1963

The Services of

MESSRS. MOORE BROS. (KELVEDON) LTD.

will be operated by

The Eastern National Omnibus Co. Ltd.

as follows :—

Service No. allocated	
38	CHELMSFORD — HATFIELD PEVEREL — MALDON
53	CHELMSFORD — COLCHESTER
70	COLCHESTER — COGGESHALL — BRAINTREE
71	COLCHESTER — EASTHORPE — KELVEDON — EARLS COLNE
72	COLCHESTER — TIPTREE — TOLLESHUNT KNIGHTS (via Black Pits)
73	COLCHESTER — TIPTREE — KELVEDON
74	COLCHESTER — TIPTREE — TOLLESHUNT KNIGHTS
331	BRAINTREE — KELVEDON — WEST MERSEA (Summer only)
332	BRAINTREE — PATTISWICK
334	TIPTREE — KELVEDON — SILVER END

TIME AND FARE TABLES WILL REMAIN UNCHANGED

REGULATIONS AND CONDITIONS. Passengers are conveyed subject to the advertised conditions contained in the current official Time Table Booklets, obtainable (Price 1/-) at any "Eastern National" Office, or in any separate notices.

THE EASTERN NATIONAL OMNIBUS CO., LTD.

Local Offices :	'Phone	Head Offices :	Local Offices :	'Phone
Braintree, Fairfield Road	104		Kelvedon, High Street	306
Chelmsford, Duke Street	3104/5	New Writtle Street,	Maldon, Omnibus Station	265
Clacton-on-Sea, Jackson Rd.	944/5	CHELMSFORD	Silver End, Boars Tye Rd.	10
Colchester, Queen St. and		'Phone 3431 (4 lines)	West Mersea, High St. ...	17
Omnibus Station	77671			

EN AD. 8279/8163 5M Index Publishers Limited, Dunstable and London—21401

Fig. 24 EN leaflet announcing they will be operating Moore's services as from 3rd February 1963.

And so after 148 years of carrying passengers Moore Bros. as a going concern was no more. All staff were transferred to The Eastern National Omnibus Company so the familiar faces would continue to be seen on the same vehicles — at least for the time being. As Harold Moore said at a press interview on the takeover, 'They've been a grand bunch and relations have been excellent. We've never had a strike throughout our history'. Such relationships contributed in no small measure to the way in which staff looked after the passengers, leaving a lasting impression on those who had travelled on a Moore's bus. With over 3½ million passengers carried over ¼ million miles during the last year of operation there would be many recalling those happy past times.

Over the next few years many changes were to take place, adversely affecting staff and passengers alike. Whilst sad to relate, this happily is not part of the Moore's story.

60 Moore's Directors and Staff from the post-war years.

Harold Moore

Percy Moore

C. Fincham, C. Mayhew
C. Goodchild, G. Worricker, J. Carrol, J. Pitchford, W. Foakes, C. Sorrell, C. Wade
G. Aves, S. Jones, C. Smith, G. Mayhew, R. Smith, A. Polley

Hubert Everitt

Bert Drane

GARAGE STAFF
C. Mayhew, F. Boyden, C. Smith, J. Gentry, H. Newport, B. Hull (Foreman), C. Smith

. . . . . . and so farewell . . .

. . . to the Moore family business.

Farewell gathering at the 'Star & Fleece'
Thursday 31st January 1963
some of the faces in the picture:-

Harold Moore	Harry Hayward
Mrs. Rene Moore	Jack Gentry
Percy Moore	Dick Greenleaf
Fred Smith	Don Rippingale
Charlie Lines	Cyril Eady
Wally Lines	George Mayhew
Reg Lines	Cyril Mayhew

Chapter 12

The Preserved Guy

61 Ordered by Moore's but delivered to Eastern National. 582AOO in Chelmsford High Street c1967.

Reference to Appendix VI will show a list of the vehicles that were taken into Eastern National ownership on 3rd February 1963 following the signing of the Agreement of Sale with Moore's on 31st December 1962. Although all the vehicles, apart from PHK533 which was in chassis form only, acquired Eastern National fleet numbers, many were withdrawn during 1963 and were up for disposal by the year's end. The list contains both the 1963 and 1964 fleet numbers together with all disposals and their subsequent history, where this is known, of each ex-Moore's vehicle.

61a 373WPU in service with Eastern National on route 38 Chelmsford - Maldon, turning into Maldon Omnibus Station.

As we have seen in Chapter 11, the last buses to be purchased by Moore's and delivered into their ownership in 1961 were two Guy Arab IV double-deckers, 372WPU and 373WPU. When these were withdrawn by Eastern National in 1974, Joe Long of the Eastern National Preservation Group, had his eye on 373WPU and so when this came up for sale in March 1975 he was a very willing buyer. Joe Long had always been a busman, first working for Westcliffe-on-Sea Motor Services and then the Eastern National Omnibus Company who, from 1952, operated the Westcliffe services on behalf of the British Transport Commission. Both Westcliffe and Eastern National had been part of Thomas Tilling Limited since the early 1930's, until that company was sold to the British Government in 1948.* So whilst Joe worked for one of the larger bus combines he nevertheless had a great interest in the independent operators of East Anglia. Thus when the opportunity arose in 1975 to purchase a bus which previously had belonged to an independent, Joe was very quick to put in a purchase bid and was delighted when it was accepted.

With the purchase of 373WPU Joe now owned one of the last two vehicles delivered new into the Moore's fleet. These two Guys were fitted

* In 1986 the assets of ENOC Ltd were acquired through a management buyout as part of the government's privatisation of the National Bus Company and its subsidiaries. Since then the name Eastern National has lived through a number of changes.

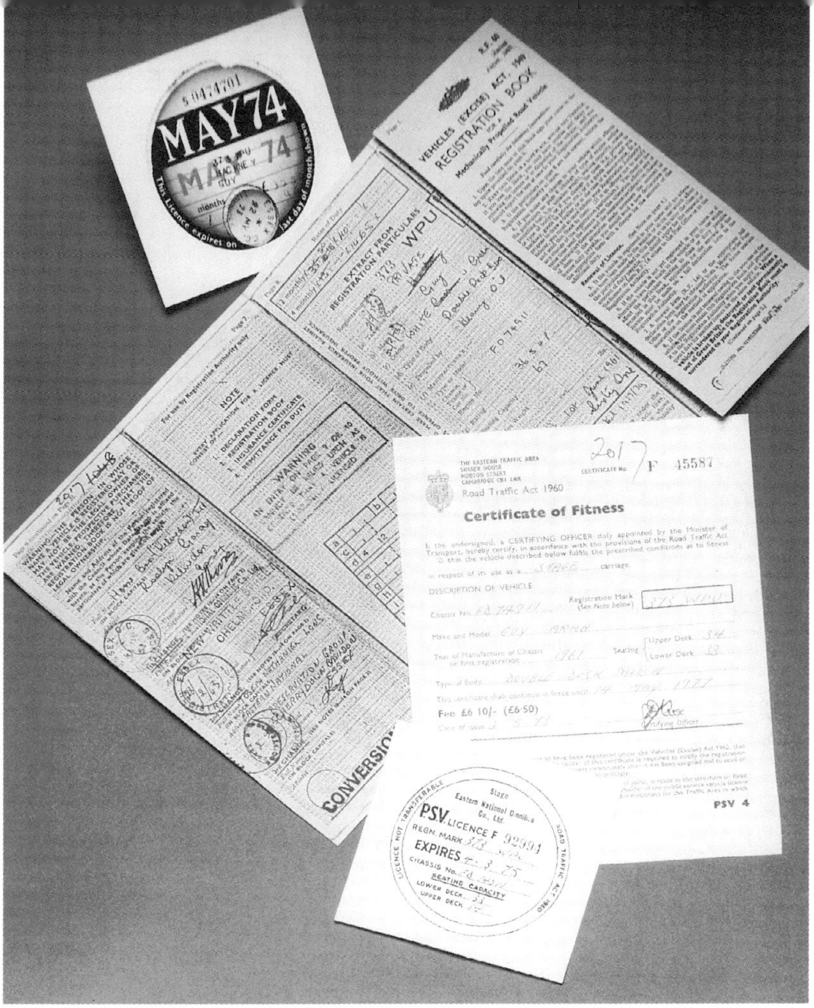

62 Last PSV documents of 373WPU.

with the so called 'Johannesburg' style concealed radiator front, regarded by many to be the most attractive produced by Guy Motors. Although the vehicle was in good condition, Joe's aim was to restore the paintwork to Moore's colours and by adding their emblem to give an air of authenticity and to his credit the results were very close to the original. As this is the only known preserved Moore's bus in the UK the following detail may be of interest. At the time of the Eastern National disposal the vehicle carried a Certificate of Fitness which was valid until 1977, but the PSV licence would expire in March 1975. However, Joe wanted the bus for private use only and arranged for an MOT, following which a Test Certificate was issued. Illustrations of the former documents together with extracts from the original Registration Book are shown in Pic. 62.

62a On display during Open Day at Canvey Bus Museum.

Joe's enthusiasm for the bus had always been evident and this grew further when past Moore's passengers, members of the Moore family and ex-staff were invited to board the bus at rallies and local events to enjoy a short period of true nostalgia. This led to Joe becoming increasingly interested in the Moore family bus business, and we know he was looking forward to the publication of this book. Perhaps his greatest thrill was the invitation by the Essex Bus Enthusiasts Group to use his bus in April 1993 over the Moore's service network to commemorate the thirtieth anniversary of Eastern National taking over Moore Bros. The bus entered Kelvedon on 3rd April in an almost carnival atmosphere with people lining the streets. At Roslyn Garage in the High Street, members of the Moore family and 27 ex-Moore's staff boarded the bus for a tour of the service routes — a memorable day for all concerned. Sadly because of feeling unwell, Joe was not at the wheel of his bus and due to his continuing ill health he was unable to enjoy the occasions when his bus took part in the Lord Mayor's 'London Parade' on New Years day in 1993 and again in 1995. During these years cancer was taking its toll which ended with his death on 4th January 1996. As a tribute to his dedication to the Canvey Island Bus Museum and his involvement in the exhibits, three double-deckers, Joe's own 1961 Guy Arab (373WPU), 1939 Bristol K (AJN825) and 1952 open-top Bristol KSW (WNO478) joined the funeral cortège on 17th January from the Trinity Methodist Church in Vange to the Pitsea New Cemetery.

Joe had made known to Brian and Arnold Brown his wish to bequeath the Guy bus to them. Brian and Arnold, who live in Basildon, had been good friends of Joe for over 20 years and so they felt honoured to receive this bequest. Before Joe's death it had been planned by all three that they would attend the 1996 International Steam Weekend in Hellevdetsluis in Holland, and almost as a form of pilgrimage Brian and Arnold drove the Guy to this weekend event in that year. In May 1997 they attended the International Motor Rally in Geraardsbergen Belgium which had been organised by the Morris Register in the UK. Later in the year they attended the annual vehicle rally in Wolverhampton where the Guy Owners Club also hold their rally. The bus and owners received a great welcome and as Brian and Arnold are members of the club further visits to Wolverhampton, which was the home of Guy Motors Limited, can be expected.

And so, the Moore's Guy has visited places in retirement which in working life would have been beyond its wildest dreams. Long may it continue and help to keep alive the name of Moore Bros., remembered with great affection by so many.

62b Moore's in the USA. 20PVX at Burbank California.

LOCAL CARRIER SERVICE — *Appendix I*

1815	1844	1865	1867	1905	1919
E. Moore	W.E. Moore	Mrs A. Moore	C.& E. Moore	Moore Bros.	

N

KELVEDON - COLCHESTER via Main Road

by 1825	by 1865	by 1878	by 1899	by 191
Wed & Sat		Mon, Wed & Sat	Wednesday	
Horse & Cart	Horse Drawn Van			

KELVEDON - CHELMSFORD via Main Road

by 1825	by 1865	by 1878	by 1917
Tues & Fri		Mon, Tues, Thur, Fri & Sat	Mo
Horse & Cart	Horse Drawn Van		

HORSE-DRAWN OMNIBUS SERVICE

1881	1905
C.& E. Moore	Moore Bros.

KELVEDON - COGGESHALL

Monday to Saturday

MOTOR BUS SERVICES (for ROUTE DETAILS see Appendix III)

Service Route Number

c1924	1926	1931 onwards	Category
2	27		
1		4	S
		19*	S
		22**	S
		27**	S
		31	S

S = Stage Carriage

* This licence surrended upon granting of No. 22

** These licences surrended upon granting of No. 31

KELVEDON - COLCHESTER via Coggeshall

1914	
Moore	
by 1921	
Wed & Sat	Mon, & S

KELVED COGGESH

THE SERVICE NETWORK

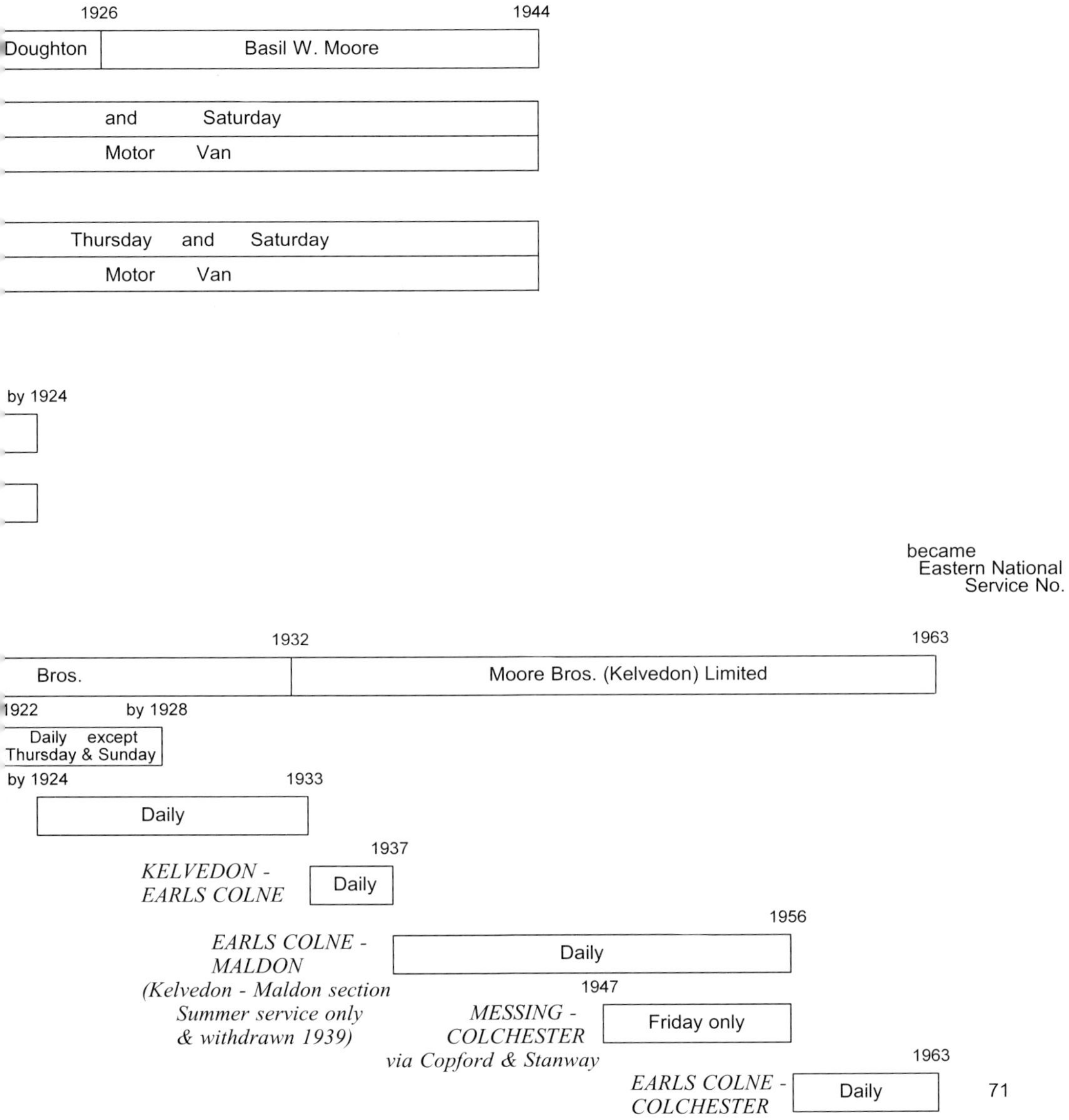

/ cont.

MOTOR BUS SERVICES (Cont)

Service Route Number

c1924	1926	1931 onwards	Category
5	27	1	S
3	19	3	S
		16	S
4		6	S
		5	S
7		7	S
6			
		2	S
		17	S
		18	S
		26	S
		30	S
		32	S
		13	E
		10	E & T
		11	E & T
		12	E & T

S = Stage Carriage

E = Express

E & T = Excursions and Tours

BRAINTRE
COLCHES

1914 | by 1921

KELVEDON - COLCHESTER
via Tiptree
** Mon, Wed & Sat one journey each direction via Messing Street*

Tuesday & Friday

1919

KELVEDON - BRAINTREE
via Halstead

KELVEDO

KELVEDON - COLCHEST
via Easthorpe

KELVEDON - COLCHEST
via Main Road

THE SERVICE NETWORK

became Eastern National Service No.

by 1924 | by 1928 | 1946 | 1963
Wed only | Daily | Hicks Services incorporated - increased frequency — 70

by 1924 | by 1928 | by 1933
xpt Sun | Daily except Sunday | * Daily | Daily — 73

TOLESHUNT KNIGHTS - COLCHESTER via Tiptree
1932
Daily — 74

1939
Wednesday only

- BRAINTREE via Bradwell
by 1928 | 1951
Daily

by 1924 | 1939
Fridays only

by 1924 | by 1928 | by 1930
Wed only | Daily expt Sun

COLCHESTER - CHELMSFORD via Main Road
1963
Daily — 53

MALDON - CHELMSFORD via Hatfield Peverel
1932
Daily — 38

BRAINTREE - WEST MERSEA via Kelvedon (Seasonal Whitsun - end Sept)
by 1933 | by 1935 | by 1938 | 1939 | 1947 | 1961
Thur Sun | Tues, Thur & Sun | Daily | Suspended | Daily | Sun only — 331

PATTISWICK - BRAINTREE via Bradwell
1947
Wednesdays only — 332

TIPTREE - SILVER END (Crittall Works)
1955
Monday - Friday — 334

TIPTREE - COLCHESTER via Haynes Green & Black Pits
1960
Mon - Fri — 72

BRAINTREE - LOWESTOFT & GT YARMOUTH (Seasonal Last Sat Jul - Last Sat Aug)
1932 | 1939 | 1947 | 1953
Sat only | Suspended | Sat only | Sat only — X71
Departure Point extended to SILVER END

From KELVEDON
1931 | 1939 | 1950
Apr to Sept | Suspended | Apr to Sept

From BRAINTREE
1946
Apr to Sept | Suspended | Licence transferred to Hicks Bros. Braintree in 1946

From COGGESHALL
Apr to Sept | Suspended | Apr to Sept

CARRIER, HORSE-DRAWN AND MOTOR BUS ROUTES

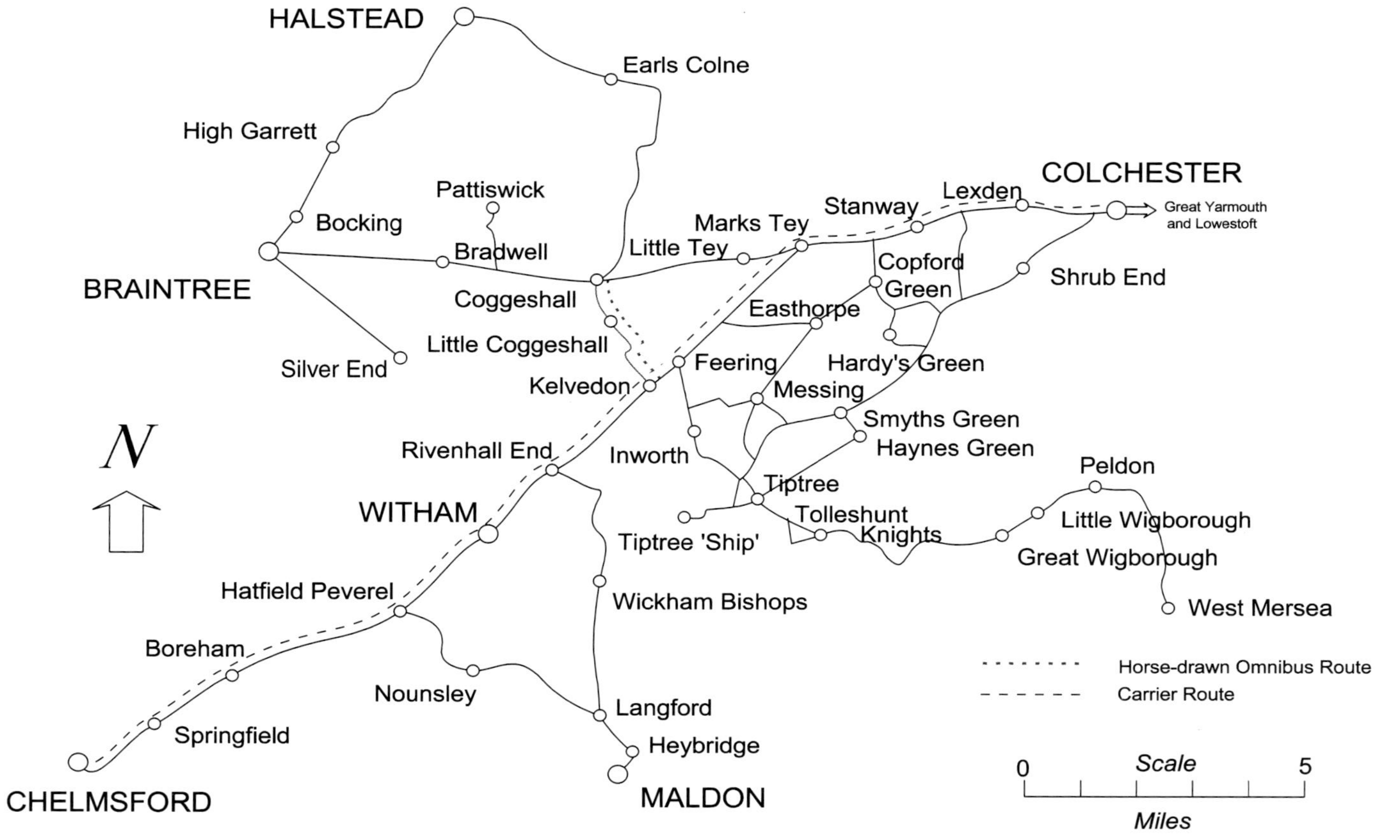

Appendix III LIST OF ROAD SERVICE LICENCES

showing Dates of Initial Applications, Grants, Refusals, Withdrawals and Surrenders
as they appeared in the Notices and Procedings 1931 - 1963

S = Stage Carriage
E & T = Excursions and Tours
E = Express Carriage

G = Grant
R = Refusal
W = Withdrawal

LICENCE No.
F/R120/Route No.

Route No.	Route	Category	Application N & P No.	Application Date	Decision	Decision N & P No.	Decision Date	Comments
1	Braintree — Colchester	S	6	27-04-31	G	38	04-01-32	Hicks Bros. licence for operation on this route surrendered to Moore Bros. Nov.1946 - see also comments Route No. 11.
2	Chelmsford — Colchester	S	6	27-04-31	G	20	10-08-31	
3	Kelvedon — Colchester via Tiptree & Messing	S	6	27-04-31	G	10	23-05-31	
4	Kelvedon — Coggeshall	S	6	27-04-31	G	16	06-07-31	It is assumed this licence was surrendered upon granting of No. 19. However no N & P record has so far been found confirming this.
5	Kelevedon — Braintree via Coggeshall	S	6	27-04-31	G	16	06-07-31	
6	Kelvedon — Braintree via Halstead	S	6	27-04-31	G	10	23-05-31	
7	Kelvedon — Colchester via Easthorpe	S	6	27-04-31	G	16	06-07-31	
8	Braintree Market Sq. to various Essex seaside towns	E & T	6	27-04-31	R	10	23-05-31	These applications were re-submitted under Nos. 10 & 11 - the reason for the Traffic Commissioner's refusal is not known. A number of other operators E & T's were refused at this time.
9	Kelvedon Garage to various Essex seaside towns	E & T	6	27-04-31	R	10	23-05-31	
10	Kelvedon Garage to:-	E & T	11	29-05-31		17	13-07-31	
	1. Clacton-on-Sea via Tiptree				G			
	2. Clacton-on-Sea direct				G			
	3. Walton-on-Naze direct				R			
	4. Walton-on-Naze via Tiptree				R			
	5. Felixstowe via Tiptree				G			
	6. Felixstowe direct				G			
	7. Southend-on-Sea via Braintree				G			
	8. Yarmouth direct				R			
	9. Yarmouth via Thetford				R			

/ cont.

Route No.	Route	Category	Application N & P No.	Application Date	Decision	Decision N & P No.	Decision Date	Comments
11	Braintree Market Sq. to:-	E & T	11	29-05-31		17	13-07-31	Licence surrendered, except sub item 8, to Hicks Bros. Nov. 1946 - see also comments Route No. 1.
	1. Clacton-on-sea direct				G			
	2. Walton-on-Naze direct				R			
	3. Felixstowe direct				G			
	4. Lowestoft direct				G			
	5. Yarmouth direct				G			
	6. Yarmouth via Thetford				G			
	7. Dovercourt direct				R			
	8. Mersea Island via Tiptree				G			
	9. Yarmouth direct period return				R			Item 9 - refusal decision appealed - again refused. Application re-submitted under Route 13 as an express service.
	10. Heybridge evening tour				G			
	11. Bulmer evening tour				G			
	12. Finchingfield evening tour				G			
12	Coggeshall Market Hill to:-	E & T	22	07-09-31		31	09-11-31	
	1. Clacton-on-Sea direct				G			
	2. Yarmouth via Thetford				G			
	3. Mersea Island half-day tour				G			
	4. Whipsnade direct				G			
13	Braintree — Yarmouth direct	E	41	25-01-32	G	51	11-04-32	
14	Braintree Market Square — Clacton-on-Sea	E	52	18-04-32	R	59	06-06-32	
15	Chelmsford Market Road — Clacton-on-Sea	E	52	18-04-32	R	59	06-06-32	
16	Tolleshunt Knights — Colchester	S	72	10-10-32	G	81	12-12-32	Originally operated by A. V. Brown Tiptree - Licence No. F/R117/1
17	Maldon — Chelmsford via Hatfield Peverel	S	72	10-10-32	G	79	28-11-32	Originally operated by A. V. Brown Tiptree - Licence No. F/R117/5
18	Braintree — West Mersea	S	106	12-06-33	G	114	14-08-33	
19	Kelvedon — Earls Colne	S	149	30-04-34	G	168	10-09-34	Licence surrendered upon granting of No. 22

/ cont.

Route No.	Route	Category	Application N & P No.	Application Date	Decision	Decision N & P No.	Decision Date	Comments
20	Hatfield Peverel — Braintree via The Notleys	S	240	27-01-36	R	248	23-03-36	Service operated by H. Springett & Son - Licence No. F/R62/1
21	Colchester — Whipsnade Zoo and Ashridge Park	E & T	296	22-02-37	G	300	22-03-37	Single occasion only - 21st July 1937, 14 vehicles permitted
22	Earls Colne — Maldon via Wickham Bishops	S	301	05-04-37	G	315	12-07-37	Licence surrendered upon granting of No. 31
23	Colchester — Felixstowe	E & T	317	26-07-37	G	320	16-08-37	Single occasion only - 26th Aug 1937, 14 vehicles permitted
24–25	These licences have not been traced							
26	Pattiswick — Braintree	S	420	25-11-46	G	424	27-01-47	
27	Messing — Colchester via Copford Green	S	420	25-11-46	G	423	13-01-47	Licence surrendered upon granting of No. 31
28	Colchester — Football Fixtures	E & T	513	03-07-50	W	519	25-09-50	
29	Braintree — Colchester Layer Rd. (Football Ground)	E	521	23-10-50	R	524	04-12-50	
30	Tiptree — Silver End (Crittall Works)	S	641	30-05-55	G	646	08-08-55	Amendment to extend to Tollesbury was refused
31	Earls Colne — Colchester via Easthorpe & Blackpits	S	653	14-11-55	G	662	19-03-56	
32	Tolleshunt Knights — Colchester via Haynes Green	S	771	23-05-60	G	782	24-10-60	

Note :

Following the transfer of services to Eastern National in 1963 the surrender of all Moore's licences appeared in N & P Nos. 847 and 849 dated 22nd April and 20th May 1963 respectively.

Appendix IV **Extracts from N&P's showing Examples of Road Service Licence Applications and Grants**

1. Stage Carriage - N&P No. 51 Date 11/04/32 RSL Grant

F./R.120/1 (Jan. 11, 1932). To **Messrs. Moore Bros.,** Roslyn Garage, Kelvedon, Essex, to continue to operate a service of stage carriages between Braintree and Colchester, via Coggeshall and Marks Tey, in accordance with their application subject to the following conditions:--

Standard conditions.

Fares to be charged shall be in accordance with the fare table as hereunder.

The times at which the service is to be operated shall be in accordance with their published time table, as deposited with the application, and subject to the seasonal modifications as detailed hereunder.

Period of operation :--Daily throughout the year.

Date of expiry :--30th November, 1932.

Date of decision :--10th March, 1932.

SEASONAL MODIFICATIONS TO TIME TABLE.

Good Friday as Sunday service.

Easter Monday as Sunday service. 10.40 p.m. ex Colchester to Coggeshall in addition.

Whit Monday as Sunday service. 10.40 p.m. ex Colchester to Coggeshall in addition.

August Bank Holiday as Sunday service. 10.40 p.m. ex Colchester to Coggeshall in addition.

Boxing Day as Sunday service. 10.40 p.m. ex Colchester to Coggeshall in addition.

Christmas Day : Dept. Braintree to Colchester, 9.0, 10.30 a.m

„ Coggeshall to Colchester, 8.20, 9.25; 10.35 a.m.

„ Colchester to Braintree, 9.0, 10.30 a.m.

„ Coggeshall to Braintree, 8.35, 9.37, 11.7 a.m.

FARE TABLE.

Braintree
2d Bartrams Avenue
2d 1d Kings Lane and The Dolphin
3d 2d 1d Bridge Hall, Bradwell Swan and Council Houses
4d 3d 2d 1d Pattiswick Turning
5d 4d 3d 2d 1d Stock Street and Holfield Grange.
7d 6d 5d 4d 3d 2d Coggeshall
8d 7d 6d 5d 4d 3d 2d Surrex
9d 8d 7d 6d 5d 4d 3d 1d Broad Green
9d 8d 7d 6d 5d 4d 3d 2d 1d Salmons Lane
10d 9d 8d 7d 6d 5d 4d 2d 2d 1d Little Tey Church
11d 10d 9d 8d 7d 6d 5d 3d 2d 2d 1d Red Lion
1/- 11d 10d 9d 8d 7d 6d 4d 3d 3d 2d 2d Marks Tey
1/1 1/- 11d 10d 9d 8d 7d 5d 4d 4d 3d 2d 1d Copford Corner
1/2 1/1 1/- 11d 10d 9d 8d 6d 5d 5d 4d 3d 2d 1d Stanway Swan
1/2 1/1 1/- 11d 10d 9d 8d 6d 5d 5d 4d 3d 3d 2d 1d Rose Gardens
1/3 1/2 1/1 1/- 11d 10d 9d 7d 6d 6d 5d 4d 3d 2d 2d 1d Villa Road
1/6 1/5 1/4 1/3 1/2 1/1 1/- 10d 9d 9d 8d 7d 6d 5d 4d 4d 3d Colchester

Return Fares.

Colchester to	Stanway Swan ..	7d	Colchester to	Coggeshall	1/6
	Copford	9d		Bradwell	2/-
	Marks Tey	10d		The Dolphin ..	2/3
	Red Lion	11d		Kings Lane ..	2/3
	Little Tey	1/-		Braintree	2/6
	Salmons Lane ..	1/1	Braintree to	Coggeshall	1/-
	Broad Green ..	1/2		Marks Tey	1/8
	Surrex	1/4		Villa Road	2/-

/cont.

SUPPLEMENTARY TO FARE TABLE.

Children under 14 years, half adult fare.

Period Tickets.

		Weekly.	Monthly.	Quarterly.
Braintree	to Colchester	.. 9/-	30/-	85/-
Coggeshall	to ,,	.. 5/-	17/6	50/-
Marks Tey	to ,,	.. 3/-	10/6	30/-
Coggeshall	to Braintree	.. 4/-	14/-	37/6
Bradwell	to ,,	.. 2/-	7/-	20/-

School Childrens' Season Tickets per School Term.

Braintree	to Colchester	.. 40/-	Coggeshall	to Braintree	.. 21/-
Coggeshall	to ,,	.. 30/-	Bradwell	to ,,	.. 13/6
Marks Tey	to ,,	.. 22/6			

2. Express Carriage - N&P No. 51 Date 11/04/32 RSL Grant

F./R.120/13 (Jan. 25, 1932). To **Messrs. Moore Bros.**, of Roslyn Garage, Kelvedon, Essex, to operate a new service of express carriages between Braintree and Yarmouth, via the following route:--

Outward--Colchester, Ipswich, Woodbridge, Saxmundham and Lowestoft.

Return--By same route.

In accordance with the time table and fare table set out hereunder and subject to the following conditions:--

Standard conditions.

Passengers in Gt. Yarmouth to be set down or picked up only at the Brewery Plain, Gt. Yarmouth.

Passengers to be taken up at Braintree, Bradwell, and Coggeshall for Lowestoft and Yarmouth only.

Period return passengers only to be carried.

Period of operation :--On Saturdays only from 30th July to 27th August inclusive.

Local conditions as to routes, parking places, etc., in Ipswich and Lowestoft.

Date of expiry :--30th November, 1932.

Date of decision :--9th March, 1932.

TIME TABLE.

(Saturdays only from July 30th to August 27th inclusive.)

		a.m.			p.m.
Braintree	dep.	7.0	Yarmouth	dep..	1.0
Bradwell	,,	7.12	Lowestoft	,,	1.30
Coggeshall	,,	7.25	Coggeshall	arr.	5.5
Lowestoft	arr.	11.0	Bradwell	,,	5.18
Yarmouth	,,	11.30	Braintree	,,	5.30

FARE TABLE

(Return Fares only)

Braintree to Yarmouth	16/-	Braintree to Lowestoft	15/-
Bradwell to ,, ...	16/-	Bradwell to ,, ...	15/-
Coggeshall to ,, ...	16/-	Coggeshall to ,, ...	15/-

All fares must be pre-booked.

/cont.

3. Excusions and Tours - N&P No. 22 Date 07/09/31 RSL Application

F./R.120/12 Application has been made by Messrs. Moore Bros., of Kelvedon, Essex, for a road service licence to run a group of excursions and tours at inclusive return fares only, starting from Market Hill, Coggeshall, as follows:--

(1) Day tour to Clacton-on-Sea, on the following route:--
By A.120 to Marks Tey.
A.12 to Colchester.
A.133 to Clacton.
Return by the same route..

(2) Day tour to Yarmouth, on the following route:--
By A.120 to Braintree.
A.1017 to High Garrett.
A.131 to Sudbury.
A.134 to Thetford.
A.11 to Norwich.
A.47 to Acle.
A.1096 to Yarmouth.
Return by A.12 to Marks Tey.
A.120 to Coggeshall.

(3) Half-day tour to Mersea Island, on the following route:--
By B.1024 to Kelvedon.
B.1023 to Tolleshunt Knights.
By-road to Salcot Cross, Great Wigborough and Peldon.
B. 1025 to Mersea.
Return by same route.

(4) Day tour to Whipsnade(Zoological Gardens), on the following route through this Traffic Area:--
By A.120 to Bishops Stortford.
A.11 to Sawbridgeworth.
A.414 to St. Albans.
A.5 to Markyate Street.
Then into the East Midlands Traffic Area.
Return by the same route.

Other excursions and tours on special occasions.
All tours will be run between April and September.
The maximum number of vehicles to be used on this group of excursions and tours on any one day is 4.

Note:

In the early part of 1931 when initial applications for licences were made and the Traffic Commissioners gave their decisions, very little detailed information appeared in the Eastern Area N&P's. It is for this reason that a Stage Carriage Grant example has been taken from 1932. However, those shown under E & T and Express Carriage, are examples of initial Applications and Grants respectively which appeared later on in 1931, and 1932.

Appendix V MOTOR FLEET 1912 - 1963

Reg. No.	Chassis Make & Type	Chassis No.	Bodybuilder	Seats	First Reg.	To Moores	Original Owner	Disposal
???	Daimler			14		1912	H. L. Mills	
F 6060	Ford T	---	*Lt Landaulet*	---	17-09-12	New		
F 7257	Ford T	2807517	*Landaulet*	---	10-06-13	New		
F 7627	Ford T	---	*Car*	4	09-08-13	06-08-15	J. Grey, Rettendon, Essex	
F 8536	Clarkson 32 h.p.	---	Hurst Nelson	O34R	26-03-14	New		1921
???	???		*Hearse*	---		1915		1925
F 5619	Thornycroft 30 h.p.	1106	*Charabanc*	14	10-05-12	27-07-15	Holmes & Smith Ltd., Southend, Essex	09-09-18
HK 2588	Ford T	1062036	*Landaulet*	---	19-01-17	New		
HK 4232	Thornycroft J	7116	---	O40R	04-02-19	15-05-19	A.W.Berry, Colchester, Essex	
HK 4696	Ford T	719332	*Touring Car*	---	21-05-19	New		1921
HK 6350	Thornycroft J	7773	ex British	O36R	14-11-19	New		
HK 7284	Ford T	3687849	*Van body with seats*		09-04-20	New		19-07-20
HK 8140	Ford T	3697798	*Touring Car*	---	21-06-20	New		
HK 8589	Ford T	---	*Charabanc*	14	22-07-20	New		
HK 8599	Ford T	---	*Bus*	14	26-07-20	New		
NO 2732	Thornycroft J	4002	---	O40R	29-06-21	New		
NO 5124	Thornycroft J	3213	ex National	O40R	03-04-22	New		
XL 5891	Daimler CB	CB2576	---	O36R	1922	New		1926
PU 2972	Ford	TT9128260	*Charabanc*	14	24-04-24	New		
PU 8901	Ford	---	*Hearse*	---	30-06-25	New		1935
PU 9308	Chevrolet	S5166T	---	B14F	21-07-25	New		
TW 8006	Chevrolet LM	15327	---	B14F	31-03-27	New		
TW 8435	Reo Sprinter	145422	---	C20F	18-04-27	New		1934
TW 9926	Reo Sprinter	146448	---	B20F	09-06-27	New		1932
VW 2275	Chevrolet LM	18275	---	B14F	17-12-27	New		1933
VW 2851	Chevrolet LO	40038	---	B14F	30-01-28	New		
VW 2852	Reo Pullman	GB163	---	B26F	30-01-28	New		1933
VW 3178	Reo Sprinter	FAX3031	---	B20F	27-02-28	New		
VW 6756	GMC T40E	404264	Economy	B26F	29-09-28	New		
VX 51	Chevrolet LQ	52578	Economy	B14F	04-05-29	New		
VX 297	Thornycroft A6	18276	---	B26F	15-05-29	New		1935
VX 298	Thornycroft A6	18298	---	B26F	16-05-29	New		1938
VX 2632	GMC T30	302815	---	B20F	16-10-29	New		1938
VX 3808	Gilford 168OT	11207	Wycombe	B32F	18-01-30	New		1939
MO 7415	Hudson Car *-registered as PSV at Eastern Area Traffic Office*							
VX 4075	Gilford 168OT	11247	Wycombe	B32F	28-02-30	New		1938
VX 4279	GMC T60D	601000	---	B32F	10-03-30	New		1937
VX 4637	GMC T60D	60924	---	B32F	24-03-30	New		1937
VX 7065	GMC T30C	307980	---	B20F	31-07-30	New		1937
PU 6622	Buick Car - *registered as PSV at Eastern Area Traffic Office*				13-03-25	1930	A. Nelson, Lexden,	1937
EV 179	Gilford 168OT	11877	Petty Bros.	C32C	30-04-31	New		1939
EV 180	Gilford 168OT	11879	Petty Bros.	C32C	30-04-31	New		1941
EV 2510	Gilford 168OT	11948	Petty Bros.	C32C	31-07-31	New		1942
EV 5299	Gilford 168OT	12046	Petty Bros.	C32C	06-04-32	New		1940
EV 5300	Gilford 168OT	12047	Petty Bros.	C32C	06-04-32	New		1940
EV 5301	Bedford WLB	108459	Petty Bros.	C20F	06-04-32	New		1937
EV 6337	AEC Regal	6621350	Petty Bros.	C32C	31-05-32	New		1954
VX 2630	Gilford 166OT	11095	Wycombe	B32F	16-10-29	1932	A. V. Brown (Bluebird) Tiptree, Essx	1937
KP 7060	Gilford 166OT	10770	Strachans	B32F	14-05-29	1932	Sheppey Motor Transpt., Sheerness, Kent	1937
UV 3200	Gilford 166OT	10843	Wycombe	C30F	09-07-29	1933	Venture, Hendon, Middlesex	1936

Note: Seats Column conforms to *PSV Circle* and *Omnibus Society* convention of:
Body Type / Seating Arrangement / Entrance Door Position

/ cont.

Reg. No.	Chassis Make & Type	Chassis No.	Bodybuilder	Seats	First Reg.	To Moores	Original Owner	Disposal
BEV 464	Albion Valiant PV70	11509J	Duple	C32F	06-06-34	New		1950
BEV 465	Albion Valiant PV70	11509H	Duple	C32F	06-06-34	New		1949
PH 1998	Chrysler	---	*Hearse*	---		1935		1950
BPU 202	Bedford WLB	109902	Duple	C20F	01-10-34	New		1945
BPU 676	Bedford WLB	109886	Duple	B20F	16-10-34	New		1945
VW 7483	GMC T42D	42479	Waveney	B26F	05-12-28	Dec 34	A.V.Brown (Bluebird), Tiptree, Essex	1936
CHK 401	Bedford WLB	110026	Duple	C20F	01-05-35	New		1959
CHK 952	AEC Regal	6621740	Duple	C32F	31-05-35	New		1953
CHK 953	Albion Valiant PV71	11602A	Duple	C32F	31-05-35	New		1953
DNO 835	Albion Victor PK115	25009A	Duple	C31F	29-05-36	New		1947
DNO 836	Albion Valkyrie PW141	44001I	Duple	C35F	29-05-36	New		1947
DNO 837	Albion Victor PK115	25010F	Duple	C31F	02-07-36	New		1947
EVW 868	Albion Victor PK115	25014L	Duple	C31F	31-05-37	New		1947
EVW 869	Albion Valkyrie PW141	44011G	Duple	C35F	31-05-37	New		1947
FEV 806	Bedford WTB	111715	Duple	C20F	23-07-37	New		1957
WL 9062	AEC Regent	661115	Hall Lewis	L24/24R	1930	1937	City of Oxford G131	1955
WL 9063	AEC Regent	661121	Hall Lewis	L24/24R	1930	1937	City of Oxford G137	1955
WL 9068	AEC Regent	661116	Hall Lewis	L24/24R	1930	1938	City of Oxford G132	1955
GNO 267	Albion Valkryie SPPW141	44017K	Duple	C32F	02-06-38	New		1954
GNO 268	Albion Valkryie SPPW141	44017L	Duple	C32F	30-06-38	New		1955
GVW 634	Bedford WTB	8470	Duple	C20F	30-11-38	New		1961
GVW 635	Bedford WTB	8988	Duple	C20F	30-12-38	New		1962
VH 4875	Leyland TD2	2244	Leyland	H24/24R	1933	1940	Hanson, Huddesfield 60	1946
UF 6468	Leyland TD1	71409	Leyland	H24/24R	1930	1940	Southdown 868	1946
WN 4762	AEC Regent	6611899	Brush	L25/26R	1932	1941	South Wales 262	1947
JPU 640	Bristol K5G	57078	Duple	L27/28R	13-07-42	New		1958
JTW 447	Guy Arab I 5LW	FD25917	Brush	L27/28R	12-05-43	New		#
JVW 999	Guy Arab II 5LW	FD26955	Roe	L27/28R	02-08-44	New		#
JVX 223	Guy Arab II 5LW	FD27478	Strachans	L27/28R	05-04-45	New		1958
JVX 553	Bedford OWB	26699	Mulliner	B32F	07-06-45	New		1951
JVX 554	Bedford OWB	26799	Mulliner	B32F	29-05-45	New		1951
JVX 555	Guy Arab II 5LW	FD27433	Strachans	L27/28R	12-07-45	New		#
JVX 556	Guy Arab II 5LW	FD27753	Strachans	L27/28R	24-07-45	New		#
JVX 557	Guy Arab II 5LW	FD27767	Strachans	L27/28R	24-07-45	New		#
JVX 558	Guy Arab II 5LW	FD27775	Strachans	L27/28R	28-08-45	New		#
LPU 611	Guy Arab III 5LW	FD28324	Strachans	L27/28R	24-03-47	New		#
LPU 612	Guy Arab III 5LW	FD28325	Strachans	L27/28R	24-03-47	New		#
LPU 940	Guy Arab III 5LW	FD29455	Duple	DP35F	12-05-47	New		#
LTW 62	Guy Arab III 5LW	FD29466	Duple	DP35F	12-05-47	New		#
LVW 124	Guy Arab III 5LW	FD29483	Duple	DP35F	12-05-47	New		#
LVW 125	Guy Arab III 5LW	FD29485	Duple	DP35F	12-05-47	New		#
ONO 537	Guy Arab III 5LW	FD36192	Strachans	L27/28R	19-07-49	New		#
ONO 538	Guy Arab III 5LW	FD36193	Strachans	L27/28R	19-07-49	New		#
PHK 531	Guy Arab III 5LW	FD70106	Strachans	C35F	31-10-49	New		#
PHK 532	Guy Arab III 5LW	FD70108	Strachans	C35F	09-11-49	New		#
PHK 533	Guy Arab III 5LW	FD70153	Strachans	C35F	29-12-49	New		#
PHK 534	Guy Arab III 5LW	FD70154	Strachans	C35F	29-12-49	New		#
PPU 29	Humber Pullman	HXO/9800446	*Hearse*	---	29-01-50	New		1955
RTW 222	Bedford OB	OB147090	Duple	C27F	31-01-51	New		#
MMT 876	Bedford OB	OB38520	Duple	C29F	1946	1951	Fallowfield & Britten, London, E. 8	Scrapped after accident in 1954
HYM 685	Bedford OB	OB51452	Duple	C29F	1947	1951	Fallowfield & Britten, London, E. 8	#
GYL 983	Guy Arab II 5LW	FD27560	Weymann	L27/28R	1945	1951	Birch Bros. K103	#
GYL 984	Guy Arab II 5LW	FD27563	Weymann	L27/28R	1945	1951	Birch Bros. K104	#
GYL 981	Guy Arab II 5LW	FD27533	Weymann	L27/28R	1945	1952	Birch Bros. K101	#
GYL 982	Guy Arab II 5LW	FD27530	Weymann	L27/28R	1945	1952	Birch Bros. K102	#

/ cont.

Reg. No.	Chassis Make & Type	Chassis No.	Bodybuilder	Seats	First Reg.	To Moores	Original Owner	Disposal
KGT 378	Guy Arab III 6DC	FD33251M	Strachans	C35F	1949	1953	Dartmouth Coaches, London, S. E. 26	#
KGT 379	Guy Arab III 6DC	FD33294M	Strachans	C35F	1949	1953	Dartmouth Coaches, London, S. E. 26	#
KGT 380	Guy Arab III 6DC	FD33307M	Strachans	C35F	1949	1953	Dartmouth Coaches, London, S. E. 26	#
KGT 381	Guy Arab III 6DC	FD33272M	Strachans	C35F	1949	1953	Dartmouth Coaches, London, S. E. 26	#
YEV 263	Guy Arab IV 5LW	FD72118	Strachans	L28/28R	27-02-54	New		#
52 DHK	Guy Arab IV 5LW	FD72582	Strachans	L28/28R	23-06-55	New		#
53 DHK	Guy Arab IV 5LW	FD72583	Strachans	L28/28R	23-06-55	New		#
576 FVW	Commer Avenger III	T83A0431	Duple	C41F	07-03-56	New		#
577 FVW	Commer Avenger III	T83A0432	Duple	C41F	07-03-56	New		#
2834 F	Guy Arab IV 5LW	FD73909	N.C.M.E.*	L31/32R	04-07-57	New		#
2835 F	Guy Arab IV 5LW	FD73910	N.C.M.E.*	L31/32R	04-07-57	New		#
8935 NO	Guy Arab IV 5LW	FD74120	Massey	L34/33R	11-09-58	New		#
8936 NO	Guy Arab IV 5LW	FD74124	Massey	L34/33R	11-09-58	New		#
19 PVX	Guy Arab IV 5LW	FD74286	Massey	L34/33R	28-06-59	New		#
20 PVX	Guy Arab IV 5LW	FD74288	Massey	L34/33R	28-06-59	New		#
7652 VW	Commer Avenger IV	94Y8063	Yeates	C41F	01-06-60	New		#
372 WPU	Guy Arab IV 5LW	FD74910	Massey	L34/33R	31-05-61	New		#
373 WPU	Guy Arab IV 5LW	FD74911	Massey	L34/33R	31-05-61	New		#

* Northern Counties Motor and Engineering Co. Ltd.

\# passed to Eastern National in 1963

Breakdown Vehicles

CHK 401	Converted from 20 seater coach to open truck (see earlier entry)	
LTW 62	Withdrawn about 1961 and converted to a towing truck	#
7654 PU	Commer Service Van --- New in 1959	#

Vehicles purchased for Spares

EO 7896	Guy Arab I	Acquired in 1954	Honeywood, Stanstead, Suffolk
OVW 756	Guy Arab III	Acquired in 1959	Eastern National, Chelmsford, Essex
OVW 757	Guy Arab III	Acquired in 1959	Eastern National, Chelmsford, Essex

Disposal Details (where known)

F 5619 - HK 4696 - HK 7284 - VW 2275 became goods vehicles

VX 3808 - VX 4279 - UV 3200 - FEV 806 - GNO 267 - GNO 268 - WL 9062 became showman vehicles

EV 6337 - BEV 465 became caravans/storesheds

WL 9068 - GVW 634 to dealer and possible scrap

VX 7065 - CHK 952 - CHK 953 scrapped

XL 5891	Tebbell, Maldon, Essex
TW 8435	Quantrill, Chelmondiston, Suffolk
TW 9962	Cracknell, Brentwood, Essex
VW 2852	Howe, Fromes Hill, Hereford
VW 3178	Theobold, Long Melford, Suffolk
VW 6756, JPU 640, JVX 223	Ashdown, Danbury, Essex
KP 7060	Ford, Althorne, Essex
VX 297	Carr, Annfield Plain, Co. Durham
VX 4075	Norfolk, Nayland, Suffolk
VX 4637	Brown (Bluebird), Tiptree, Essex
EV 179	A. & E. Pettifer, Bromyard, Worcs
EV 180, DNO 837, EVW 868	Tye Bros., Mendlesham,
VX 2630, EV 2510, EVW 869	Allen, Reepham, Norfolk
EV 5299	Rayner (P & M), Ipswich, Suffolk
EV 5301	Rule, Boxford, Suffolk
BEV 464	Wiffen, Finchingfield, Essex
BPU 202	Rolfe, Hillingdon, Middlesex
BPU 676	Lodge, High Easter, Essex
DNO 835	Everson, Wix, Essex
DNO 836	Jolly, Norton, Suffolk
VH 4875	Kearsey, Cheltenham, Glos.
UF 6468	Skinner, Hastings, Sussex
WN 4762	London Brick Co.,Stewartby, Beds.
JVX 553, JVX 554	Morley's, West Row, Suffolk

Appendix VI LIST OF VEHICLES TRANSFERRED TO EASTERN NATIONAL AND THEIR SUBSEQUENT DISPOSAL

Reg. No.	Eastern National Fleet Nos. 1963	1964	Owner	Date
HYM 685	025	—	1. Twell (dealer), Ingham, Lincs	Oct 1963
			2. C. E. Smith, Ingham, Lincs	Dec 1963
			Withdrawn	1966
RTW 222	026	—	1. Twell (dealer), Ingham, Lincs	Oct 1963
			2. Winton (Contractor)., Lough, Lincs	Dec 1963
			3. J. Stainton, Alvington, Lough, Lincs	Dec 1967
			Later scrapped	
LPU 940	027	—		
LVW 124	028	—	1. Twell (dealer), Ingham, Lincs	Oct 1963
LVW 125	029	—		
KGT 378	030	—	1. Twell (dealer), Ingham, Lincs	Oct 1963
			2. North Ray Farm Products, North Thoresby, Lincs	May 1964
			3. Twell (dealer), Ingham, Lincs	Oct 1964
KGT 381	031	—	1. Twell (dealer), Ingham, Lincs	Oct 1963
			2. Kirkby (dealer), South Anston, Yorks	Mar 1964
KGT 379	032	—	1. Twell (dealer), Ingham, Lincs	Oct 1963
			2. B. Holland (showman), South Wigston, Leics	Jun 1964
KGT 380	033	—	1. Twell (dealer), Ingham, Lincs	Oct 1963
PHK 531	034	—	1. Twell (dealer), Ingham, Lincs	Oct 1963
			2. Kirkby (dealer), South Anston, Yorks	Mar 1964
			3. Dorman Long Ltd., Middlesborough, Cleveland	Mar 1964
			Later scrapped	
PHK 532	035	—	1. Twell (dealer), Ingham, Lincs	Oct 1963
PHK 534	036	—	1. Twell (dealer), Ingham, Lincs	Oct 1963
			2. Henry Hodges Ltd., Sheffield, Yorks	Sep 1964
PHK 533	—	—	Chassis only - believed scrapped	
576 FVW	037	100	1. Wicks, Braintree, Essex	Aug 1965
			2. PVS (dealer), Silver End, Essex	Jun 1972
			3. United Pentecostal Church of God, London, E5	Oct 1972
			Withdrawn	1978
577 FVW	038	101	1. Wicks, Braintree, Essex	Sep 1965
			2. Shaw and Kilburn (dealer), London	Sep 1973
			3. R. W. Denyer, Stondon Massey, Essex (not licenced)	Unknown
			4. W. J. Sheahy, London, E8 (caravan use)	Mar 1975
			5. C. J. Thomson, London, SW19 (caravan use)	Aprl 1975
			Scrapped	Mar 1977
7652 VW	039	102	1. Twell (dealer), Ingham, Lincs	Sep 1968
			2. William Varah (Transport) Ltd., Wickersley, Yorks	Sep 1968
			3. Mick Smaller's Luxury Coaches, Barton-upon-Humber	Apr 1972
			Withdrawn	Jan 1973

/ cont.

Reg. No.	Eastern National Fleet Nos. 1963	1964	Owner	Date
LTW 62	040	—	1. Twell (dealer), Ingham, Lincs	1964
JTW 447	1024	2000	1. Twell (dealer), Ingham, Lincs	Oct 1966
			2. G. Armitage, Ardsley, Maltby, Yorks	May 1967
			3. ??? (dealer), Batley, Yorks	Aug 1967
JVW 999	1025	2001	1. Twell (dealer), Ingham, Lincs	Oct 1966
			2. C. A. Whiteley Ltd., Ardsley, Maltby, Yorks	Oct 1966
			3. Geest Industries Ltd., Spalding, Lincs	Nov 1969
JVX 555	1026	—	1. Eastern National Preservation Group, Canvey Island, Essex - vehicle used as a store Later scrapped	Mar 1974
GYL 982	1027	2002	1. Twell (dealer), Ingham, Lincs	Oct 1965
GYL 981	1028	2003	2. Mick Smaller's Luxury Coaches, Barton-upon-Humber	Oct 1965
			3. Twell (dealer), Ingham, Lincs	Jun 1966
GYL 983	1029	2004	1. Twell (dealer), Ingham, Lincs	Oct 1966
GYL 984	1030	2005	1. Twell (dealer), Ingham, Lincs	Sep 1965
			2. Redland Transport Ltd., Reigate, Surrey (based at Yaxley, Cambs)	Oct 1965
			3. Mick Smaller's Luxury Coaches, Barton-upon-Humber (not operated)	Jul 1969
			4. Twell (dealer), Ingham, Lincs	Jul 1969
JVX 556	1031	—	1. Twell (dealer), Ingham, Lincs	Dec 1963
JVX 557	1032	—	1. Twell (dealer), Ingham, Lincs	Oct 1963
JVX 558	1033	—	1. Twell (dealer), Ingham, Lincs	Dec 1963
LPU 611	1034	2006	1. Twell (dealer), Ingham, Lincs	Oct 1966
			2. C. A. Whiteley Ltd., Ardsley, Maltby, Yorks	Oct 1966
			3. Mellers Coaches Ltd., Goxhill, Barton-upon-Humber	Dec 1970
			Withdrawn	Dec 1971
LPU 612	1035	—	1. Twell (dealer), Ingham, Lincs	Oct 1963
ONO 537	1036	—	1. Audawn Coaches, Corringham, Essex	Mar 1964
ONO 538	1037	—	2. Bone (dealer), Highwood, Essex	Feb 1965
			3. Mountnessing Auto Spares, Mountnessing, Essex Believed scrapped	1966
YEV 263	1038	2007	1. Twell (dealer), Ingham, Lincs	Oct 1967
			2. Mick Smaller's Luxury Coaches, Barton-upon-Humber	Nov 1967
			Withdrawn	Jun 1968
			3. Twell (dealer), Ingham, Lincs	Feb 1969
52 DHK	1039	2008	1. Twell (dealer), Ingham, Lincs	Oct 1967
			2. Mick Smaller's Luxury Coaches, Barton-upon-Humber	Nov 1967
			Withdrawn	Nov 1968
			3. Twell (dealer), Ingham, Lincs	Feb 1969

/ cont.

Reg. No.	Eastern National Fleet Nos. 1963	1964	Owner	Date
53 DHK	1040	2009	1. Twell (dealer), Ingham, Lincs	Oct 1967
			2. Mick Smaller's Luxury Coaches, Barton-upon-Humber	Dec 1967
			Withdrawn	Jul 1969
2834 F	1041	2010	1. Twell (dealer), Ingham, Lincs	Sep 1969
2835 F	1042	2011	2. C. A. Whiteley Ltd., Ardsley, Maltby, Yorks	Oct 1969
			3. Twell (dealer), Ingham, Lincs	Oct 1969
			4. Mick Smaller's Luxury Coaches, Barton-upon-Humber	May 1971
			Withdrawn	May 1972
8935 NO	1043	2012	1. Twell (dealer), Ingham, Lincs	Feb 1971
			2. C. A. Whiteley Ltd., Ardsley, Maltby, Yorks	Apr 1971
			3. H. Jarvis (Coaches), Maltby, Yorks	Jan 1975
			Withdrawn	Jun 1976
8936 NO	1044	2013	1. Twell (dealer), Ingham, Lincs	Feb 1971
			2. C. A. Whiteley Ltd., Ardsley, Maltby, Yorks	Apr 1971
			3. H. Jarvis (Coaches), Maltby, Yorks	Jan 1975
			Withdrawn	Oct 1976
19 PVX	1045	2014	1. North (dealer), Sherburn, Yorks	May 1975
			2. Ensign (dealer), Hornchurch, Essex	Jun 1975
			3. Ensign Bus Company Ltd., Hornchurch, Essex	Jun 1975
			4. Exported to Holland, on behalf of Grey-Green, but still owned by Ensign for use on publicity. Returned to UK	Apr 1976 Jun 1977
			5. Becket (dealer), Carlton, Yorks	Sep 1977
20 PVX	1046	2015	1. North (dealer), Sherburn, Yorks	May 1975
			2. Ensign (dealer), Hornchurch, Essex	Jun 1975
			3. Ensign Bus Company Ltd., Hornchurch, Essex	Jun 1975
			4. Exported to RWE Swank and Associates, Orange, California, USA	Jul 1976
			5. Butchers Arms public house, Burbank, California	Oct 1980
			6. London Bus Company, Burbank, California	Dec 1982
			7. Los Angeles Motor Coach Company, Marina del Rey, California	May 1997
372 WPU	1047	2016	1. North (dealer), Sherburn, Yorks	May 1975
			2. Ensign (dealer), Hornchurch, Essex	Jun 1975
			3. Robin Hood Coaches, Ilford, Essex	Aug 1975
			4. Ensign (dealer), Hornchurch, Essex	Apr 1976
			5. Don Lee, General Contractor, Santa Ana, California, USA	Apr 1976
373 WPU	1048	2017	1. Joe Long, Basildon, Essex (Eastern National Preservation Group, Bus Museum, Canvey Island, Essex)	Mar 1975
			2. Brian and Arnold Brown (ENPG), Basildon, Essex	Jan 1996
581 AOO	1049	2018	1. P. Sykes (dealer), Blacker Hill, Yorks	Nov 1975
582 AOO	1050	2019		
7654 PU	Commer Van			